HOW
TO
STAY
ALIVE

For more information on Bear Grylls and his books,
see his website at **www.beargrylls.com**

Bear Grylls

HOW

🔥

TO

✈

STAY

➕

ALIVE

☢

THE ULTIMATE SURVIVAL GUIDE
FOR ANY SITUATION

wm

WILLIAM MORROW
An Imprint of HarperCollinsPublishers

Thank you to the BG team of survival and safety experts I have worked with on so many adventures, especially to Scott, Stani, Meg and Dave for all your input to the practical, resourceful and innovative details that have helped me compile this book. I dedicate it to this team, who have lived and breathed all this stuff alongside me for so many years and in so many hellholes!

HarperCollins books may be purchased for educational, business, or sales promotional use. For information, please email the Special Markets Department at SPsales@harpercollins.com.

FIRST PUBLISHED IN GREAT BRITAIN IN 2017 BY BANTAM PRESS, AN IMPRINT OF TRANSWORLD PUBLISHERS

FIRST WILLIAM MORROW PAPERBACK EDITION PUBLISHED 2018.

Photograph of George Mallory standing near Everest on p. 275 printed by courtesy of the heirs of George Ingle Finch.

Designed by Julia Lloyd

Library of Congress Cataloging-in-Publication Data has been applied for.

ISBN 978-0-06-285711-8

HB 01.23.2023

CONTENTS

GREAT ESCAPES

HOW TO...

TERRAIN SURVIVAL

HOW TO...

LIFE-OR-DEATH SITUATIONS

HOW TO...

MEDICAL EMERGENCIES

HOW TO...

INTRODUCTION

I have spent much of my life travelling to some of the most inhospitable places on the planet. I've summitted Everest, traversed the treacherous Northwest Passage in a small inflatable boat, and survived some of the most remote and wild deserts, jungles and swamps on earth.

Along the way, I've found myself in some pretty hairy situations, moments when my own life, and those of my fellow travellers, hung in the balance. *How to Stay Alive* is the distillation of the survival wisdom I have accumulated over the years. During my time in the British Special Forces, I applied many of these survival principles firsthand.

I begin *How to Stay Alive* with the basics, the rudiments of any adventurer's education: what you need to carry in your survival kit, how to make fire and purify water, and how to tie the three most important knots that will help you get out of a jam. Then we get into more exotic territory. You may have never found yourself in a sandstorm, but I can tell you that you'll want to be prepared. Likewise apocalyptic snowstorms, flash floods and tornadoes.

There are other books out there about survival, but these pages hold the tactics and techniques that go far beyond bushcraft. We cover it all, from kidnappings to car brake failures, from shark attacks to how to fly a plane in an emergency. So dig in— and the next time you find yourself lost, cornered or in a spot of trouble, you'll be properly prepared.

SURVIVAL SKILLS

SKILLS

THE

BASICS

HOW TO

ASSEMBLE

A

LIFE-SAVING

BACKPACK

In the military, you get used to carrying heavy backpacks filled with over 100lb of gear. It gets you fit and it gets you strong. But in a survival situation, lugging too much stuff around with you could be a killer. It slows you down and it drains you of energy.

So now, whenever I go out on an expedition, I take the very minimum I can get away with. With a light pack on your back, you can move with speed and agility over rugged terrain. You can beat the weather if a storm's coming in. You can make sure that your energy is directed towards the important business of getting yourself out alive, rather than being sapped by useless pounds of excess weight you really don't need.

KEEP IT DRY

Before thinking about what we need to put in our backpack, we need to make sure it's waterproof. I don't care if a rucksack manufacturer claims that their rucksack is 100 per cent waterproof: they never are. Not in the kind of conditions you might encounter. No rucksack will withstand wading across a lagoon or fording an underground river.

So you always need a liner of some sort. If it's a proper rucksack liner, great. If not, a plastic bag will do.

Inside that liner, you need . . . another liner! One bag is never enough in water. Soldiers regularly put dry clothes in double bags. If it's something really important like a radio then it's even more crucial to double-line it. (I've lost count of the number of people I've met who have bought a 'waterproof bag' in which to stow their phones, only to find that it's not waterproof at all. Two bags is *always* the way to go.)

Once you've waterproofed your rucksack, you can think about what to stow in it.

ESSENTIAL GEAR

A KNIFE
Carry a knife, save a life. See pages 59–63 for what you need to know.

EMERGENCY RATIONS
At the bottom of your bag, you should always carry some emergency rations. A good stash would be a bag of nuts, energy bars and flapjacks. Put them in a small bag and wrap it all up tightly with masking tape so it's like a solid brick. Stow it away at the bottom of your pack and forget about it, safe in the knowledge that if a day-long expedition turns into an overnighter, you've got a solid lump of a thousand or so calories sitting there ready for use.

A WATER FILTRATION SURVIVAL STRAW

These straws are plastic tubes with built-in filters. You simply put one end of the straw into contaminated water – be it in a bottle, a river, or even a puddle – and suck clean water through the top of the straw. The filter does all the hard work for you. I often carry one of these: they're cheap, light and effective. Good bits of kit.

WARM CLOTHING

Clothing can be heavy and bulky. You don't want to be carrying too much. The really essential piece of gear is a warm, waterproof jacket to give you that extra layer of protection against the elements. Try to choose one with waterproof zips, Velcro cuffs and an elasticated hem. Above all, it needs to be lightweight.

A MAP

If you venture out into unfamiliar terrain without a map, you're asking for trouble.

A COMPASS

On pages 71–8 you'll find lots of tips about how to find your way when you're lost using the sun, the stars and other forms of natural navigation. In terms of equipment, however, your best friend is always going to be a compass. It doesn't need to be big, expensive or fancy – in fact, you want it to be small and light.

A 2-METRE LENGTH OF CORD

This has so many uses. For example:

- building shelters (see pages 47–50)
- fishing (see pages 99–105)
- trapping (see pages 107–113)
- kit repair
- as a boot lace
- as an emergency tourniquet (see page 414)

A HEAD TORCH

Even better than a head torch? Two head torches. And a couple of spare batteries. Loads of people get themselves into dangerous situations because they slightly misjudge their timings and end up in unfamiliar terrain with no light source. A head torch shows you the way and keeps your hands free.

A OAN OΓ OΓɅAY ΓLAOTCΓ

You only need a small one. It's antiseptic as well as being able to seal small cuts or grazes.

SURVIVAL TIN AND FIRST AID KIT

See the next section for what these need to contain.

And that's it. Lightweight, but containing all the items you really need.

K.I.S.S.
(KEEP IT SIMPLE, STUPID!)

Keep it waterproof. You need *two* waterproof liners.

Keep it light. Heavy backpacks slow you down and drain your energy.

Emergency rations will last for years.

Waterproof jacket. Map. Compass. Cordage. Head torch. Job done.

HOW TO

PUT

TOGETHER

THE PERFECT

SURVIVAL

KIT

My survival kit and my first aid kit go together. I have a little waterproof bag to carry both inside my backpack. For both kits, I prioritize portability. There's no room for any luxuries – I only want to be carrying stuff around that I absolutely might need in a survival situation.

FIRST AID KIT

Don't get me wrong: when you're out with the family, a little first aid kit containing a few plasters and some creams for bites and the like is fine.

I take the view that I can probably make do with a few cuts and grazes (unless I'm in the jungle, when I take them super-seriously: don't underestimate the importance of regularly cleaning small cuts in such an environment to avoid infection). In most situations I want stuff that's going to save my life or someone else's, and which doesn't weigh a ton. So this is what I take.

GAFFER TAPE

You know the stuff – thick, black, sticks to anything. It has loads of uses. You can use it to strap people's feet together if they have a broken bone (see page 434). You can use it to close a wound that requires stitches (see page 425). You can stick it over sucking chest wounds to aid breathing and stem bleeding (see page 413). You can use it in place of an eye patch over a damaged eye. In Afghanistan, soldiers would stick it over bullet wounds. Gaffer tape is a million times more useful than plasters. If you fall over on to a stick and puncture your lung, a plaster's a waste of time. A bit of gaffer tape might just give you a chance.

QUIKCLOT

Quikclot is a product used in hospitals by first responders and in the military to deal with catastrophic bleeding injuries. Originally it was a powder that you poured into a wound to promote blood clotting. The version I carry is a gauze, which weighs nothing but punches way above that. You stuff the gauze into the wound – be it an animal bite, a severed arm or a bullet wound – to stem the bleeding and clot the blood quickly.

A FIELD DRESSING
To stem bleeding from traumatic wounds.

OROPHARYNGEAL AIRWAY
This is a plastic tube which you place in a patient's mouth to help

keep their airway open. It does this by stopping the tongue from covering the epiglottis, which often happens when somebody becomes unconscious. (They're not used when patients are conscious because they can stimulate the gag reflex.)

You could also take a nasal airway. These fulfil a similar function, but are inserted into the nostril rather than the mouth.

ASHERMAN CHEST SEAL

This is a special dressing for traumatic chest wounds with a one-way valve that allows blood and air to escape, but doesn't let either back into the chest cavity.

TOURNIQUET

Standard first aid advice is to avoid the use of tourniquets. In extreme environments, however, and in situations where you have a catastrophic bleed that *has* to be stopped, tourniquets save hundreds of lives (see pages 414–15 for more on this). I carry a standard military tourniquet, but belts or boot laces will do the job too.

EPIPEN

I'm highly allergic to bee stings, so I carry one of these. It gives me a shot of adrenalin to fight life-threatening anaphylactic shock. Know your weaknesses and you stay strong.

ZINC OXIDE TAPE

Once you've read the section on how to prevent blisters (see pages 397–400) you'll understand that this is the stuff you need.

SURVIVAL KIT

Everything in my survival kit is stashed in a military tobacco tin, sealed with gaffer tape. The tin keeps everything dry, safe and ready for emergencies, but it's also a useful item in its own right. For example, it can be used as a container to heat up water, or to make tinder for the following night's fire. To do this, tear off a piece of cotton from your T-shirt. Place it in the empty tobacco tin and put the tin over the fire. You'll find that the cloth goes black and rigid. It's known as 'char cloth'. Stash it away safely and when you go to light your fire the following night, the char cloth will be a life-saving piece of tinder and will light immediately.

Inside my tobacco tin, I keep the following items.

WATERPROOF MATCHES

Lighters are unreliable in cold temperatures and in the wet. Waterproof matches are your best bet for fire-starting in difficult conditions.

A 9-VOLT BATTERY AND SOME STEEL WOOL

A good back-up to waterproof matches. See page 38 for the low-down on how to use these items to get a fire going.

BUTTON COMPASS

These are tiny, cheap and weigh nothing. See page 72 for more.

WATER PURIFYING TABLETS

On pages 25–31 you'll find some tips on collecting and purifying water, but tablets are the easiest and quickest way of making sure your life-saving water is safe to drink.

A TEA LIGHT

Caves, forests at night . . . there are many locations and times when you might need a bit of sustained light that doesn't require a battery. Tea lights are cheap, weigh very little and fulfil this purpose. They are particularly useful in Arctic and Antarctic environments. If you've had to dig yourself into a snow hole (see pages 51–2), a single candle reflecting off the walls of a white cave will provide a lot of light. Perhaps more importantly, you can use it as an indicator of the level of oxygen in the atmosphere. If it starts guttering, that's a sign you need to make a hole in the roof to let some more air in.

If you can find candles made of tallow – which is animal fat – you can also eat them. Not the best meal you'll ever have, but ready calories in a survival situation.

A CONDOM

I take these mainly for carrying water. A condom takes up almost no space but is very elastic and can hold up to two litres of water. Since they're waterproof, they're also good for keeping tinder dry. And you can use them as an improvised rubber glove if you're treating a wound and want to guard against infection from a dirty hand. Choose a non-lubricated version.

A TAMPON
For firelighting: the absorbent material inside a tampon makes good tinder.

K.I.S.S.
(KEEP IT SIMPLE, STUPID!)

Your first aid kit and survival kit need to be small and light.

Focus on what you need in an emergency, not on what will make you more comfortable.

Gaffer tape is your friend.

A military tobacco tin is more than just a container.

HOW TO

COLLECT

WATER

AND MAKE

IT SAFE

Water is gold dust. You can survive for a couple of weeks without food, but only a couple of days without water – and even less than that in the desert. In a survival situation, pretty much nothing else in this book matters if you don't have access to clean drinking water.

'Clean' is the operative word. Water-borne pathogens are some of the biggest killers in the world. That aside, the main problem with drinking dirty water in a survival situation is that it will likely make you vomit and/or give you diarrhoea. If this happens, you'll lose more water than you take in. When Aron Ralston famously got trapped under a boulder in the Utah desert and ended up having to amputate his own arm, he also drank from a puddle of filthy, stagnant water (not to mention having previously had to make do with his own pee) when he was walking across the desert to safety. Within minutes his bowels expelled the fluid. Drinking that water hindered him by making him even more dehydrated through diarrhoea.

Your chances of staying alive are almost always improved if you know how to collect water and make it safe to drink. Here are the main methods of water collection in the wild, and some instructions on quickly getting it to a drinkable state.

COLLECTING WATER

RIVERS, STREAMS AND LAKES

These are your most obvious sources of water, provided you clean and purify anything you collect. Two rules for collecting water from rivers and streams:

1. If you are going to risk drinking direct from the water source, the safest water is that which is moving fast over rocks. Try to avoid collecting water from slow-moving pools.
2. If you're collecting river or stream water in a bottle, don't let the opening face upstream – this will funnel in all the small debris, twigs and the like. Turn it the other way round.

GYPSY WELL

If you come across a bog of dodgy, stagnant, very dirty water, you can dig a hole next to it, about a foot away and a foot deep. This is known as a gypsy well. The dirty water will seep into the hole through the adjoining earth, which will filter it to a certain extent.

You'll still need to filter and purify this water to make it completely safe to drink (see page 29), but the gypsy well will do a lot of the work for you and make your purifying methods more efficient.

If you can't purify your water, this is a way of reducing the potential hazards of drinking very stagnant water.

You can also dig a well like this in a dry river bed. If there's any sub-surface water, it should seep into the hole.

ABOVE-GROUND SOLAR STILLS

When green leaves photosynthesize, they give off oxygen and water vapour. You can collect this water by wrapping a plastic bag around a branch of green, non-poisonous vegetation. Tie the mouth of the bag tightly and leave it for several hours. You can get a good inch of water in your bag if you do this right. The more bags you have, the more water you can collect.

BELOW-GROUND SOLAR STILLS

These are especially good in the desert, where you'll have few other ways of collecting water. Dig a hole about 60cm deep and maybe a metre across. Put a container in the bottom, then lay a piece of plastic sheeting over the hole. Put rocks all around it to hold it in place, then put a stone in the middle, right above the container, so the plastic sheet forms a cone shape pointing downward. Leave for several hours: moisture from the ground will condense on the underside of the sheet and drip into the container.

If you urinate in the hole before setting up the still, the moisture in your urine will condense into clean drinking water.

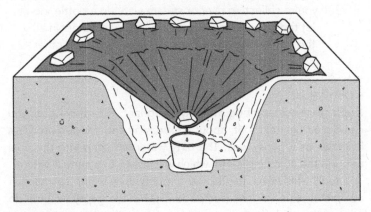

DEW COLLECTION

Where morning dew has collected on grass or other vegetation, you've lucked out because it's easy to collect. Wrap towels, rags or any absorbent clothing round your feet, then walk through the vegetation. When the fabric has become saturated, you can wring it into a container and repeat the process until you've collected as much water as you can.

MOSS

This acts like a sponge. And like a sponge, you can squeeze damp moss to produce small quantities of water.

FILTERING WATER

A Millbank bag is a constituent part of many military belt kits. It's a fabric bag into which you pour your unfiltered water. Any debris gets caught inside, while cleaner water drips through the fabric. Worth having, but you can improvise a similar process.

Take off your sock. Better still, take off your underpants. Even better, use a pair of tights. Pour your unfiltered water through the fabric so that the worst of any debris is filtered out.

This is the quickest, easiest way to filter water in a survival situation. You can make a better job of it by filling your sock or whatever you're using with sand and rocks, layering it with the least coarse material at the bottom and the more coarse material at the top. But if you need to get water into your body fast, a rough filter will probably suffice.

Alternatively, carry a water filtration survival straw (see page 16).

PURIFYING WATER

I always carry purifying tablets in my survival kit. If my water is not too dirty I'll chuck one in. If it's honking, I'll chuck in two or three. It won't taste great, but it'll do the job.

Alternatively you can boil your water. People will tell you that you need to boil it for ten minutes. If you've got an unlimited water source, that's fine. If you've collected only a limited amount of water, the more you boil it, the more you lose in evaporation. So once it's come to the boil, drink it: 99.9 per cent of the pathogens will have been killed.

PRESERVING THE WATER IN YOUR BODY

You can reduce the amount of water you need to take in by reducing the amount of water you excrete. So:

1. As far as possible, stay out of the sun and in the shade.
2. Keep out of the wind: it evaporates moisture from your skin and causes you to sweat more.
3. Eat less. Your body uses water to help digest food.
4. Keep your mouth shut and breathe through your nose – you lose much more water vapour through your mouth.
5. Move less and more efficiently – see the section on conserving energy (pages 53–7).
6. If you come across undrinkable water, use it to dampen your skin and clothes and so cool your body down, which will mean less sweating.
7. Urinate less – the longer your body can hold on to that water, the better.

URINE

I know, I know . . . but urine is a life-saver if you have no other source of water. It's almost sterile when fresh, but you don't want to drink it if it's very dark: it'll poison you rather than hydrate you.

SEA WATER

Don't go there. See pages 188–9 for water collection tips at sea.

K.I.S.S.
(KEEP IT SIMPLE, STUPID!)

In a survival situation, water = life.
Prioritize it above everything else.

Contaminated water can make you excrete more than you drink:
weigh up the risks and benefits before drinking it.

Collect. Filter. Purify.

Preserve the water in your body: take steps to sweat
and urinate less, and keep your mouth shut.

HOW

TO

MAKE

FIRE

Fire keeps us alive. We can use it to boil water and make it safe to drink. To cook food. To keep warm. To ward off dangerous animals. To send emergency signals. To make tools. In a survival situation, fire is your friend.

Many people, however, find it difficult to light a fire in a grate at home, let alone when their life depends on it. That's because they don't understand fire: what it is, what it needs and how to make it.

WHAT IS FIRE?

Fire is what we see and feel when a particular type of chemical reaction, called combustion, occurs. Let's not get too sciencey, but combustion happens when oxygen reacts with some sort of fuel, which must be a gas. In order for this to happen, solid fuel must be heated up to the point where it becomes gaseous. This is its 'ignition temperature'. Once that happens, the reaction will continue to provide more heat, but if the temperature drops, the reaction will stop.

WHAT FIRE NEEDS

So the three things a fire needs to burn are the three elements of the combustion reaction: oxygen, fuel and heat. We call this the fire triangle. It is easy to remember as humans require the same three elements to survive.

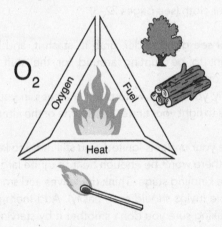

If you remove any of these three elements, the fire will go out. The more oxygen there is, the faster and hotter a fire will burn.

HOW TO MAKE A FIRE

The secret of making a good fire lies in remembering the fire triangle, and understanding that combustion won't occur until you've raised your fuel to the correct temperature. Imagine trying to ignite a log using only a match. It won't burn, because the match doesn't supply enough heat to get the log up to temperature. If you use the same match to light a piece of dry grass, however, the grass *will* burn.

So, in order to start a fire, we need material that will combust with a tiny flame or even a spark. This is called tinder. Good materials for tinder include:

- dry grass
- pine cones
- birch bark
- cotton wool
- the insides of tampons
- empty birds' nests
- char cloth (see pages 22–3)

If you see good tinder, grab it, stash it, and keep it dry. You don't want to be hunting around for the stuff when your life depends on it.

Ideally you'll have waterproof matches in your survival pack. Use these to light your tinder, or use one of the alternative methods below.

Once your tinder is ignited, you still need to keep things small because there won't be enough heat to ignite large pieces of fuel. This is the kindling stage. Think dry leaves and small twigs – as dry as possible (twigs should snap easily). Add them gradually to the flame, making sure you don't smother it by starving it of oxygen.

Only when you have a decent blaze should you start adding larger pieces of fuel. But do it gradually; never rush things. Remember the old adage: look after a fire when it's small, and it will look after you when it's big.

Having said that, keep it small. In general, small fires are best because:

- they're easy to manage
- they use less fuel
- they're less likely to get out of control
- in most situations, a small fire will perform all the life-saving functions as successfully as a big one

PRIOR PLANNING . . .

. . . and preparation prevent poor performance, as the old army saying goes. In most survival situations, you're living on your wits, reacting fast. But with fire-making you have to think ahead and gather all your materials before you try to start one. Otherwise your fire will go out while you're hunting for the right fuel, and it can't perform its life-saving function.

So: preparation, preparation, preparation. Make several piles of different-size fuel before you start.

I always think that the appropriate type of wood for a fire when snapped should sound like the crackle of a fire. This is a good indicator that the wood is dry and dead rather than green, wet and living. Living wood won't crackle when snapped.

WHAT IF IT'S WET?

First, build a raft of wood as a platform to keep your fuel off damp ground.

You can start a fire with wet wood, but it's obviously more difficult. Your initial heat needs to be greater, which means that in this instance a larger fire is better. As the ground is likely to be wet, make a bed for your fire out of fallen logs first. Using your knife, you can also shave wet, dead wood down to get to the dry inner layers of the log.

OTHER TYPES OF FUEL

Wood isn't always available. I'm not saying the following options are going to give you the most pleasant fire you've ever built, but when your life depends on heat and warmth, you might need to improvise. Try:

- dried animal droppings – these can be mixed with dry grass to make a good fire
- engine oil or petrol/diesel scavenged from a vehicle (see pages 167–8)
- spare car tyres, or upholstery scavenged from a vehicle (see page 168)
- dry grass, bundled together to make logs
- dead animals – if you've eaten the meat and the animal fat is going to go rancid, you can make a raised bed of animal bones, and place the fat on top and some tinder underneath to make a fire

STARTING A FIRE WITHOUT MATCHES

Sounds tough, but it doesn't have to be. Waterproof matches are always in my survival kit (see page 23), but if you find yourself without them, here are a few ideas for improvising.

1. **Use a 9-volt battery and some steel wool.** I always carry this in my survival kit as a back-up for my waterproof matches. Touch the steel wool to the contacts of the battery. The electric charge will cause the steel wool to burn.

2. **Use your mobile phone.** This is potentially very dangerous, but most modern phones have a lithium battery. If your mobile is dead or broken, you can break into it and remove that battery. If the lithium inside is exposed to air or water, it will burn – or even explode, so this should only be attempted as a last resort. Use a knife to hack into the side of the battery and have some tinder ready to burn as the lithium ignites.

3. **Use a lens.** As every schoolkid knows, a magnifying glass will concentrate the sun's rays, and this will be enough to light dry tinder. You can also use the bottom of a glass bottle, or a piece of ice carved into a convex shape and polished.

4. **Use a liquid lens**. A clear sandwich bag, a balloon or even

a condom, filled with water until they're spherical, will focus sunlight as well. You'll need to hold them just a couple of inches from your tinder. Of course, try not to drip water on to it.

5. **Use the jump leads from a car, and a pencil.** Carefully connect the leads to your car battery, then clip them to either end of the pencil. It should catch fire.

6. **Use a flint and steel.** You can buy special fire strikers that will give a spark no matter what the weather, but a piece of flint struck against the blade of a steel knife will do the job as well.

K.I.S.S.
(KEEP IT SIMPLE, STUPID!)

Remember the fire triangle: oxygen, fuel, heat.
Lose any of these, you lose the fire.

Tinder. Kindling. Fuel. In that order.

Start slow and stay patient: time spent
building a fire is never wasted.

Small is better.

HOW TO

EXTINGUISH

A FIRE

You're never going to extinguish a blazing forest fire (see pages 333–8 on how to escape one), and escaping a burning building is one of the toughest survival situations there is (see pages 339–44). But these are by no means the most common causes of death and personal injury from fire. The most recent statistics available report that in one year, in the UK alone, there were 263 fire-related deaths and 7,569 fire-related casualties. Half of accidental fires are domestic, caused by cooking appliances in the home.

So, out-of-control fires are a very real danger, indoors and outdoors. If you're in the presence of one, and it truly is out of control, you need to get out of there. Don't be a hero. Fires kill quickly. Your only real chance of survival in a fire situation is to get yourself away from the heat and – crucially – the smoke.

Sometimes, though, we have a split second in which to make a potentially life-saving call, and we can kill a fire before it kills us.

Fortunately, it's easy to put out a campfire, and most public buildings and some homes have fire extinguishers – though not every extinguisher is the same. Certain extinguishers make certain fires worse. Here's how to deal with them.

TYPES OF FIRE, TYPES OF EXTINGUISHER

Fires are divided into six categories. A fire extinguisher will tell you on its side which type of fire it is suitable for. The six main types of fire extinguisher are water, dry powder, foam, CO_2, metal fire and wet chemical. Extinguishers are normally red, but each has an identifying block of colour on its body. The colours are:

- Water: red
- Dry powder and metal: blue
- Foam: cream
- CO_2: black
- Wet chemical: yellow

CLASS A: ORDINARY COMBUSTIBLE FIRES

The most common type. These occur when ordinary combustible materials – paper, wood, textiles, rubber, plastics – catch light and burn.

Class A fires can be extinguished with water, foam, dry powder or wet chemical extinguishers.

CLASS B: FLAMMABLE LIQUIDS

Think petrol. Flammable liquids only need a spark to ignite and are very difficult to firefight. You can't add water to these fires because it will make the liquids splash and spread.

Class B fires should be tackled with a foam or a dry powder extinguisher.

CLASS C: FLAMMABLE GASES

Such as propane, butane and methane. Perhaps the most dangerous fire to try to extinguish because there's often a source that's expelling the gas constantly. So the most important thing to do – if it's possible without putting yourself in more danger – is to switch off the gas supply.

The only fire extinguisher suitable for Class C fires is a dry powder one.

CLASS D: METAL FIRES

It takes a lot to ignite metal, but powdered metal and shavings are easier to set fire to. Some alkali metals react with air and water. Put them in contact with water or foam and they'll burn hotter or even explode.

There are special Class D fire extinguishers available. These are the only ones you should use.

CLASS E: ELECTRICAL FIRES

This is actually not an official UK fire class, but fires can easily be started by dodgy wiring, short circuits and faulty equipment. Electricity is not a fuel: it's an ignition source. It needs to be switched off, if at all possible, before you tackle the fire. Water and foam conduct electricity, so they shouldn't go anywhere near an electrical fire.

Electrical fires can be put out using a CO_2 or a dry powder fire extinguisher.

CLASS F: COOKING OIL FIRES

Incredibly common. Think chip pans. Many injuries are caused by people who think they can put out a chip pan fire with water. You can't. Adding water will make the fire spread very quickly, and probably harm you.

Cooking oil fires can only be tackled using a wet chemical fire extinguisher.

WHAT IF YOU DON'T HAVE AN EXTINGUISHER?

Remember, fire needs three things to burn: fuel, heat and oxygen. If you remove any of these three, you'll kill the fire.

Here are some tips:

- Water will remove heat from the fire triangle if it's a simple combustible fire. But you must never use it on flammable liquids, electrical fires or cooking oil fires.
- Smothering a fire is normally your best call. Chip fat fires should be extinguished by placing a lid or a damp cloth over the pan. Fire blankets are good to have around.

- Always kill the power if you have an electrical fire, or the gas if you have a gas fire.

HOW TO PUT OUT A PERSON
WHOSE CLOTHES ARE ON FIRE

When this happens, you have to move fast. You won't have time to fetch water, so you need to smother the flames. Your aim is to get rid of the oxygen supply. Do this by:

- rolling them around on the ground
- covering them in sand or mud
- smothering them with a blanket or towel (a wet one is best, but a dry one will do)

If it's you that's on fire, you need to follow the instructions given to schoolchildren in the US: Stop, Drop and Roll. Stop where you are; drop on to the ground and cover your face with your hands; and roll until the flames are out.

If someone's clothes have caught fire, they might be badly burned (see pages 428–9 for how to treat burns). Don't try to remove clothing that is stuck to the skin.

PUTTING OUT A CAMPFIRE

Even a small fire can be a big hazard if it's not properly extinguished. It's no good simply chucking water or sand over it, dabbing it with your hand and checking it's cool. I guarantee that if that's all you do, I could come along the next day, dig down a little, find a glowing ember and turn it into fire with my breath. And if my breath can do it, so can the wind.

So: once you've put the fire out, you need to break it apart. Get your foot in there, kick it around a bit like you see cowboys doing in the movies. Now add more water and sand. Make sure your ash has no hidden, burning depths. If water is limited, pee on it.

K.I.S.S.

(KEEP IT SIMPLE, STUPID!)

If a fire is truly out of hand, get away from it.

Check the side of a fire extinguisher to ensure
it's suitable for the fire in question.

Never add water to flammable liquids,
electrical fires or cooking oil fires.

Person on fire: roll and smother.
Self on fire: stop, drop and roll.

HOW TO
MAKE A
SURVIVAL
SHELTER

I'm going to show you how to make two very simple 'stay-alive' temporary shelters. Before I do that, let's make one thing clear: there are loads of elaborate shelters you can learn how to make. But a shelter doesn't need to be elaborate to keep you alive.

In a survival situation, it just needs to work.

In practice, this means making a shelter out of anything you can get your hands on. An old tarp. A bit of sheet metal. The only real requirements are that your shelter should keep you protected from the elements and animals. If it keeps you out of harm's way and protects you from the wind, rain, sun or snow, then it's doing its job.

Moreover, in a survival situation you certainly don't want to spend too much energy or too many hours building an unnecessarily fancy shelter. An hour, tops. You're probably only going to be there for a night. Long enough to rest, keep sheltered while you attend to your other survival priorities, and then get going again.

Unless you are making a camp for the long haul, in the hope of some form of rescue finding you (or if you have injured members with you who are unable to move), your main focus will be getting out to safety and civilization. So don't waste precious energy building what will only be a temporary structure. This is real survival, not the Swiss Family Robinson.

TWO RULES

Whether you're making a wigwam or just using a piece of tarp, remember these two rules:

1. Your entrance should face away from the wind. That is, you want your back to the wind. This means you can have a fire without the smoke blowing back into your shelter. And, of course, you won't have horrendous weather conditions breaching the shelter.
2. If you build a fire – and chances are you'll want to – situate it about a metre from the opening of the shelter and build a horseshoe shape around the back of it from rocks or whatever material comes to hand. This will make it more efficient by deflecting the heat back towards the shelter.

TWO SHELTERS

Easiest and quickest first.

A LEAN-TO

You can make a lean-to out of pretty much anything. A tarp tied between two trees and pegged into the ground will do the job. You can make something sturdier out of branches and vine (or string, cord or rope if you have it).

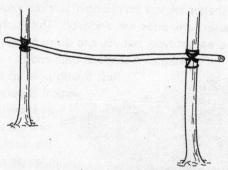

Lash a long horizontal branch between two upright trunks.

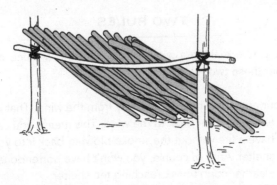

Gather more branches to lean against the horizontal one.

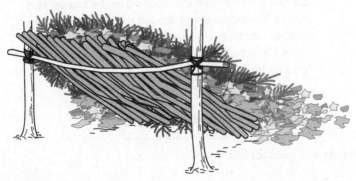

Cover the lean-to with leaves, earth or moss to waterproof it. Laying more branches or twigs over the leaves will stop them blowing away.

AN A-FRAME

An A-frame shelter takes a little longer but can provide better shelter because both sides are enclosed. This is the quickest and easiest type, but you can do something similar by making an A shape at both ends.

Rest a long pole up against two vertical support branches as shown. Use more branches to create the sides, then cover with leaves and moss as with the lean-to shelter.

SNOW CAVE

To make a shelter in the snow, you have to think a little smarter. Building a snow hole – or snow cave – is a great way of keeping your ambient temperature above zero. And again, it need not be a lot of work: keep it simple. If you're stranded in the snow it will save your life.

Dig into a bank of deep drift snow (these gather mainly on windward slopes where the snow gets piled up). Use a shovel if you have one, or improvise using anything that will scoop snow efficiently. Use your hands if you have nothing else. Remember also to take off some layers of clothing as you will begin to sweat, which in turn will make you colder once you stop working.

Remove the scooped-out snow to one side. Tunnel in and slightly up. (Cold air will sink in your shelter so you want to be above that.) Make the entrance the lowest point.

Clear enough headroom to either sit reclined upright or lie down. Remember to insulate where you sit or lie using your backpack or some fir branches if there are any nearby. Inside the snow cave, smooth the ceiling off to stop it forming drips.

Use the snow that you have removed to make a windbreak outside your entrance. Close your entrance up as much as you can but ensure you have some ventilation coming through from

Air vent

somewhere, otherwise you will suffocate. An air hole in the ceiling, made with a ski pole or long stick, will do the job. Regularly check that this remains clear, as in a heavy snowstorm it can get covered very fast.

Don't underestimate what an amazing life-saving tool the snow hole is. A friend of mine was caught alone on the mighty Mount Elbrus in Russia. He had zero visibility and couldn't even tell which way was down. It was –30°C. He had no sleeping bag. He dug himself into a snow cave and survived in there for thirty-six hours with just a small window looking out into the storm. Only when there was a fifteen-minute break in the clouds, at four a.m., did he get a visual on the lights of base camp. He whipped out his compass, and nineteen hours later made it down. But without the snow cave, he'd have been dead.

K.I.S.S.
(KEEP IT SIMPLE, STUPID!)

Shelters don't need to be fancy. They just need to work.

Use anything you have to hand to keep the elements off.

Entrance away from the wind.

A lean-to and an A-frame are the simplest shelters in forested areas.

Snow caves save lives.

HOW TO
CONSERVE
ENERGY

Your body is like a machine. If you fuel it, it'll keep on working. If you don't, it'll break down. In a survival situation, the fuel your body needs can be in short supply. The more energy you expend, the more fuel you use. So the conservation of energy is a critical factor in keeping you alive.

Trouble is, when you go into crisis mode, your body responds in an energy-intensive way. Think increased heart rate, hyperventilation, adrenalin production. It's the fight-or-flight response, and it can drain you of energy. You need to find ways of effectively managing your body's energy use.

Here are some ways of conserving energy in the field.

BREATHE

If you can control your breathing, you can control your body's natural response to a crisis situation. Slow, deep breaths, in through your nose and out through your mouth, will reduce your heart rate and control your body's production of adrenalin and endorphins.

If you're panicking, acknowledge it to yourself. Once you know what's happening in your head, you can start to control it. By doing that, you can alter what's happening in your body.

MAKE A PLAN AND WORK ON IT FAST

Your energy levels are going to be highest at the beginning of a survival situation. This is when you're at your most effective. Try to put your panic to one side, make a plan of campaign and act on it while you have the physical strength and mental clarity to do so.

MOVE SLOW TO MOVE FAST

When you're in a dangerous environment, you want to get through it and out of it as quickly as possible. Running at top speed and pushing your body to the point of exhaustion is seldom the answer.

If you're sweating, you're using too much energy. It means your body is too hot and is having to use energy to cool you down. I often witness people making this mistake when in the wild. The best strategy is to move slowly, without sweating but without stopping. It's known as Alpine pace. There might be people moving much faster than you, and overtaking. But invariably they'll have to stop further along the trail because they're exhausting themselves, and overheating. Nine times out of ten you'll reach your goal before they do. We can learn from this in a survival situation: much better to move slowly and constantly than in short, energy-hungry bursts.

If you don't *have* to move, don't.

MAINTAIN A STABLE BODY TEMPERATURE

Even when you're not moving, you need to think carefully about your body temperature. Shivering uses up energy just as sweating does. Seek out shade when it's sunny, or cool down when you find water. When it's cold, get a fire going, build a shelter or wrap yourself in whatever comes to hand.

PICK YOUR ROUTE CAREFULLY

Whether you're in a survival situation or not, you need to be as sure as you can about the route you are taking. Time spent working this out is never time wasted. Read the sections on what to do when you're lost (pages 137–41) and how to navigate (pages 71–8), and don't expend precious energy going off on little recces if you can possibly avoid it.

Adopt a 'measure twice, cut once' mentality to route-finding and navigation. Keep checking, even when you're sure you're on the right path. The wild can play tricks on you, and being lost is an energy-drain.

TRAVEL LIGHT

If you give yourself a 100-litre rucksack, you'll cram it full of stuff you don't need. Give yourself a 30-litre rucksack, you'll force yourself to be more efficient. And in the fight for survival, efficiency is key. Bottom line: the bigger your pack, the more energy you'll use. Check out the opening section on what you really need in your rucksack (see pages 13–17), and always choose the most lightweight clothing and footwear. In a crisis, you'll thank yourself for it.

If you find yourself in a survival situation carrying too much gear, consider ditching anything unnecessary.

MAKE A MUNCH BAG

You can last for days without food, but that doesn't mean you'll be performing at your best. When you exert yourself, your body uses the most readily available energy, which comes from sugar and other carbohydrates. It'll then move on to the reserves of protein and fat, but that's a slow process and less suited for situations where you have high energy needs. If you don't replenish your sugars, you risk becoming hypoglycaemic – dizzy, nauseous and lethargic.

To avoid this, I sometimes make a munch bag. It's a military thing: a plastic bag filled with broken-up chocolate, flapjacks, dried fruits, nuts, raisins – anything that will provide an instant energy hit. I keep it in a pocket, and give myself a mouthful every twenty minutes so that my body's energy levels are kept topped up.

Even if you don't have a munch bag, remember that it's better to eat whatever food you do come by little and often, rather than in one burst.

K.I.S.S.
(KEEP IT SIMPLE, STUPID!)

Move slow to move fast.

Don't sweat, don't shiver.

Pick your route carefully.

Travel light.

Eat little and often.

HOW TO

USE

A

KNIFE

A good knife is essential. If you were to take only one piece of kit on an outdoor expedition, it would be your knife. There's an old saying: carry a knife, save a life. I'd be lost without mine.

Of course, they can also be fatal instruments if improperly used, so there are a few basic rules you need to remember.

1. Avoid the triangle of death: knee to groin to knee. Too many people use their knife to work on something that's resting on their lap. That's in easy reach of your femoral artery. If you slip and cut into that, your blood will drain out of your body in a matter of minutes, and it will be a horrible, painful way to die. So, rest anything you're going to cut on the ground, *not* on your lap.

2. Make sure there's a safe circle of space all around you when you're using your knife.

3. Don't walk around with a knife: sheathe it. But . . .

4. When you're unsheathing your knife, *never* wrap one hand round the sheath. A sharp knife can easily cut through its edges. If that happens, and you're clutching it, it can just as easily cut through the tendons in your fingers. This happens all too often with old or tatty sheaths.

5. If you're handing a knife to someone: handle first, blade upward.

6. A sharp knife is a safe knife (see below).

CHOOSING A KNIFE

There are a million survival knives out there. Here's what you need to look out for.

- A fixed blade – stronger and more reliable than a folding blade.
- The right size. A 10-inch knife with a 5-inch blade is big enough to split wood, but not too big for finer work.
- A full tang. That means that the blade and handle are made out of one piece of solid metal which goes the whole length of the knife. Stronger.
- A pointed tip, for stabbing.
- A robust pommel – the bottom bit of the handle. This means the knife can double as a hammer.

SHARPENING A KNIFE

As mentioned, a knife needs to be sharp. If it's blunt it's dangerous because you end up having to use too much pressure, and without the ability to bite properly, the blade will slip. A rule of thumb: if the blade is too blunt to cut through paper, it's too blunt to cut through wood.

You can buy all sorts of knife sharpeners and whetstones, and if you're sharpening a knife at home, these are fine.

In a survival situation, however, you're unlikely to have all that stuff. But that's not a problem, because actually *anything* can sharpen a knife, so long as it can provide a bit of friction. Even a piece of cardboard will provide some friction. It'll take longer to sharpen, but it'll do the job.

In the field, a wet rock makes a good improvised sharpener. Sharpen the blade using small circular movements. If you can set up a sequence of increasingly smooth rocks, you'll get a better hone.

And don't wait for a knife to go blunt. Sharpen it little and often. It's not over the top to spend a few seconds sharpening it every time you use it.

USING YOUR KNIFE IN THE FIELD

A good, sharp survival knife is for more than just cutting up food. Here are a few ways you might find yourself using it in a survival situation.

1. **Splitting wood.** Even a small survival knife is a perfectly good substitute for an axe in many situations. Place the sharp edge at the end of a piece of wood and pound the flat edge to cut through it. This is a good technique for whittling down larger bits of wood to make smaller pieces of firewood. If you need to cut wood to make a shelter, your knife is essential.

2. **Digging.** Use your knife to dig out edible tubers from the ground, to dig a fire pit, and to dig a hole to dispose of human faeces.

3. **As a stake.** Driven into the ground, your knife will serve as a good anchor.

4. **For hunting.** Lash your knife to a straight pole to make a spear.

5. **First aid.** Used carefully, a knife will remove splinters and pierce blisters (see page 400). You can also use it to tear cloth for makeshift bandages.

6. **As a sap tap.** Find the point where the branch of a coniferous tree breaks off. There will be a knot there. Bang your knife into it and sap will start to dribble down the blade. That sap is survival gold: it's great for fire-starting, makes decent glue, and with some varieties of tree it is a good, sweet, edible food.

K.I.S.S.
(KEEP IT SIMPLE, STUPID!)

Avoid the triangle of death.

Sharp is safe, blunt is dangerous.

There are multiple uses for a knife beyond simple cutting.

Anything that provides friction can sharpen a knife.

HOW TO

TIE

ESSENTIAL

KNOTS

If you read the knotter's bible *The Ashley Book of Knots* you'll find nearly four thousand different ones. And I'm sure that for the knot enthusiast there's always the right knot for the right job.

Very few people, though, have the time or the patience to memorize a zillion knots. And you don't have to. The truth is, there isn't much you can't do with an overhand knot, a bowline and a clove hitch.

And if in doubt, remember: if you can't tie knots, tie lots!

Having just a handful of knots that you can recall under pressure will help you in a survival situation. In this section are my four favourites. Master these, and you'll be well prepared. Knot-tying is one of the few bits of survival know-how you can easily learn and practise at home. I'd urge you to do just that. The best time to master the bowline, for example, is *before* you need it to pull someone from a ledge to the top of a cliff in an emergency situation. Practise them until they're engrained in your muscle memory and you can do it in the dark – one day you might have to do just that.

If, having mastered these four, you want to get deep into the art of knot-tying, go for it. But for most emergency situations, these are the ones you want.

OVERHAND KNOT

You almost certainly know how to tie this knot already. The overhand knot is what most of us would automatically do if we were handed a piece of string and asked to tie a basic knot. It's fundamental. If you tie it around something, it can be easily undone. If you don't, it's a jamming knot, which means it is very difficult to undo. It's a 'stopper knot', which means that it's useful for stopping the end of a piece of rope sliding through a hole. It's also good to tie it at the end of a rope to stop it from fraying. If you go any further into knot lore, you'll see that the overhand is used in the construction of many other knots.

To tie an overhand knot: make a loop, pass the leading end of the rope back through it, and pull tight. Note that when this knot comes under load it can be very hard to untie.

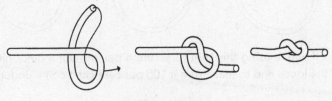

BOWLINE

This is my personal favourite, and probably the most useful knot you'll ever learn. It's an essential knot for mountaineers, as it makes a loop at the end of a rope that won't slip or tighten. It can be tied very quickly, and is probably the best way of making a lifeline to get around someone's midriff. (This makes it the ultimate survival knot: it's why all firemen are taught how to tie it.)

1.   **2.**   **3.**

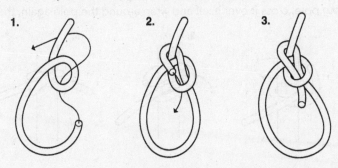

The Scouts use a great mnemonic to remember how to tie it: 'The rabbit comes out of the hole, goes around the tree and back down the hole.'

Bowlines can also be tied one-handed – a neat trick if you're hanging off a cliff.

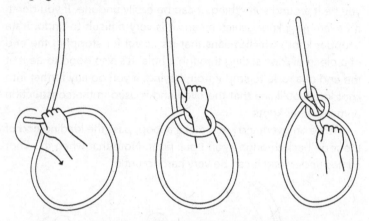

If you're using this knot to secure a person, put a quick hitch in the loose end to make sure it 100 per cent can't come undone.

CLOVE HITCH

This is the best, simplest knot for tying one end of a rope to a vertical pole, like a tree. It's a useful emergency climbing knot, as you can use it to tie the middle of a rope to a fixed anchor, and to secure yourself at the top of a climb.

To tie a clove hitch: wrap the leading end of your rope around your pole, cross it over itself and wrap around the pole again, then

1.   **2.**   **3.**

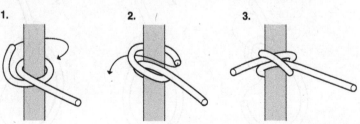

tuck the leading end back under before closing up and pulling tight. (Again, if your life depends on it, put in an extra hitch for security.)

HIGHWAYMAN'S HITCH

So called because highwaymen used to use this knot to secure their horses. It's used to secure a load that you subsequently need to release quickly. Use this knot to moor a small boat or raft, or indeed to tether an animal.

To tie a highwayman's hitch, bend the leading end of the rope back on itself. Make a second loop further up the rope and pass that through the first one. Pull the long end of the rope a little to hold the knot in place. Make a third loop in the leading end, pass it through the second and tighten again by pulling on the long end. The hitch will now hold with tension when the long end is pulled, and it will release when the leading end is pulled.

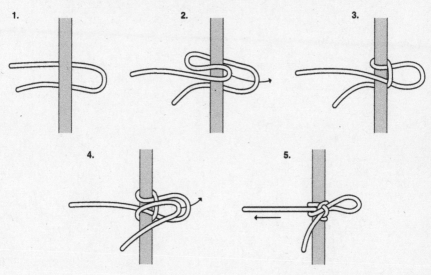

K.I.S.S.
(KEEP IT SIMPLE, STUPID!)

Forget about learning thousands of knots. In a survival situation, the best knot is the one you can tie quickly without thinking.

If you can't tie knots, tie lots.

Don't be a snob about the simple overhand knot.

Practise these knots till they're second nature – *before* you need to use them.

HOW TO

BECOME A

NAVIGATION

NINJA

GPS is awesome. The military use it all the time and it's built into almost every phone on the planet.

Awesome, but fallible. Phones break. Batteries die. Satellites fail. Heavy clouds come in, or you're under canopy. Suddenly your life-saving GPS is of zero use. If you can't navigate your way out of a life-threatening situation, you die. So you'd better have a back-up.

If you read the opening section on what you really need to stow in your backpack, you'll know that a compass is a priority (see page 16). Being able to establish directions is the core navigation skill. You don't need a flash one. A button compass will cost you fifty pence and is a life-saving piece of kit – there's a reason it's the one item many Royal Marines hide behind their cap badge.

In the absence of a compass, however, there are loads of different ways to orientate yourself. You don't need to know them all. Here are a few killer ones I always remember.

NATURAL NAVIGATION

TREES

You can gain an incredible amount of information simply by looking at a tree.

Trees love the sun. So, if you look at a tree from a distance, you'll see that the side that gets the most sun will tend to be bushier. In the northern hemisphere, that's the south side. In the southern hemisphere, it's the north side. Similarly, if you look at two trees next to each other, the tree that sees the most sun will have more growth.

MOSS

There's an old wives' tale about moss. Since moss grows in the shade, then the side of a tree that is most mossy is the northern side (in the northern hemisphere). There is some truth to this technique, but be aware that if you go into a forest, you'll see moss all over lots of trees.

Note that it's critical which tree you look at. A suitable tree needs to be:

- standing alone so it's not shaded by other trees
- vertical – if it's leaning over, damp will form on the underside, which will encourage moss and lichen anyway

If you find such a tree, ignore its bottom metre – most trees have moss around the bottom because they draw moisture from the ground. But if your tree ticks all the other boxes, look about a third of the way up. If you have growth on one side of the tree, that will point north in the northern hemisphere and south in the southern hemisphere.

PREVAILING WINDS

In parts of the world such as Africa, where there are lots of birds and plenty of predators, you'll find that birds will make five or six nests – one real nest for their eggs, the rest of them decoys. One tree can have hundreds of nests, and they'll all tend to be on one side. The birds are smart: they make their nests on the side of the tree that is sheltered from the prevailing wind. So, if the prevailing wind comes from the east, the nests will be on the western side.

Prevailing winds also have a substantial effect on the landscape. The prevailing wind in the UK is south-westerly. If you go to Dartmoor, for example, you'll see trees and even bits of gorse permanently bent over from being battered by the wind. This is a phenomenon you'll find all over the world.

So, make a point of knowing before you set out which way the prevailing wind comes from, as well as the forecasted wind direction.

THE SUN

There are two easy methods of using the sun to navigate.

THE WATCH METHOD
In the northern hemisphere, line the hour hand of your watch up with the sun. The bearing that bisects the hour hand and twelve o'clock is due south.

In the southern hemisphere, line twelve o'clock up with the sun. The bearing that bisects twelve o'clock and the hour hand is due north.

THE STICK METHOD
Put a long, straight stick in the ground in full sun. It will cast a shadow. Place a rock at the end of the shadow. Now wait fifteen minutes. The shadow will move. Place a rock at the end of the new shadow. A line joining the two rocks is east–west, the first rock being west.

ASTRONAVIGATION

Astronavigation, or navigating by the stars, really works. One of my great heroes is the explorer Ernest Shackleton. In 1915, stranded in the Southern Ocean, he navigated in a small wooden boat across 800 miles of open sea. Using the stars and dead reckoning, he was heading for the tiny island of South Georgia, and he hit it dead on.

If you look at the night sky, you'll see that the stars appear to move. Really, of course, it's the movement of the Earth that gives this effect, but it means that if you're going to use the stars to navigate, you'll need to find a fixed point. In the northern hemisphere, this means locating Polaris, the north star. In the southern hemisphere, you want the Southern Cross.

Expert navigators might know the rest of the night sky like the back of their hands, but you can get away with four constellations. Fair warning: finding these constellations can be tricky. In your back garden, there's a good chance your vision will be compromised by light pollution. In a survival situation, when you're in the middle of nowhere and far from any kind of light pollution, you have a different problem: the sky can be a blanket of pinpricks, making it hard to distinguish anything. That's why it's good to have a few different techniques up your sleeve.

FINDING POLARIS

Polaris is *not* the brightest star in the sky. But if you know how to find it, and you are in the northern hemisphere, then you will always know which way is north.

You can find Polaris using one of two constellations: the Plough (or Big Dipper) or Cassiopeia.

The Plough

The Plough has a characteristic 'saucepan' shape. Look for the two stars on the non-handle side. The line between them points to Polaris. It's about four times the distance between them.

Cassiopeia

Cassiopeia is shaped like a 'W'. It circles Polaris almost exactly opposite the Plough. If you follow a line from the centre of the 'W', you'll find Polaris, which is about equidistant from the two constellations.

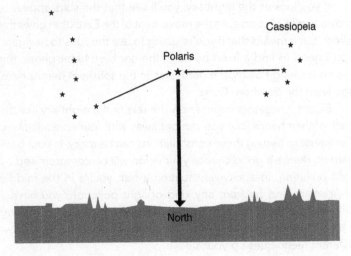

The Plough

Cassiopeia

Polaris

North

ORION

Wherever you are in the world, in the winter months Orion rises in the east and sets in the west. The most accurate part of the constellation to mark east and west is the three stars that form Orion's Belt. If you can locate them close to the horizon, you're sorted.

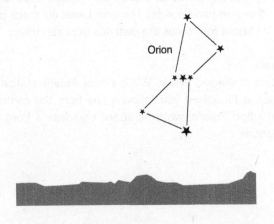

Orion

THE SOUTHERN CROSS

Polaris is a fixed point in the northern hemisphere, but it has no equivalent in the southern hemisphere. To find south, you need to locate the Southern Cross. It looks like this.

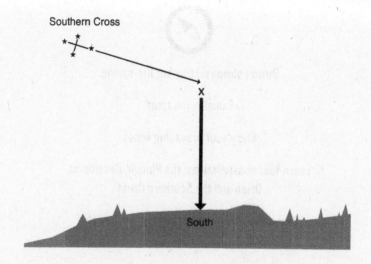

Imagine a line along the longest axis. Extend this for five times its length, then make like Shackleton and drop a vertical to the horizon. That's south.

THE MOON

The moon is visible in both hemispheres and can be used to find approximate bearings. If you have a crescent moon, draw a line between the two 'horns' and follow it down to the horizon. In the northern hemisphere, this is approximately south. In the southern hemisphere it's approximately north. The higher the moon is in the sky, the more accurate this method.

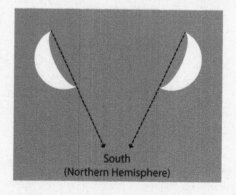

K.I.S.S.

(KEEP IT SIMPLE, STUPID!)

Button compass: tiny but life-saving.

Examine the trees.

Know your prevailing winds.

Learn four constellations: the Plough, Cassiopeia, Orion and the Southern Cross.

Remember the sun and crescent moon tricks.

HOW TO

JUDGE

DISTANCE

AND TIME

ON LAND

AND SEA

Knowing how far away an object or landmark is can be a crucial piece of information. On land, you might be deciding whether you have enough time to make for a particular position before nightfall. On water, you might be deciding whether you have the energy to swim for a point on the horizon.

Judging distance is notoriously difficult, however. Objects can appear closer or further away according to the environment or the terrain across which you're looking.

Objects appear closer when:

- the sun is bright
- the sun is behind you
- you are looking up
- the object you're looking at is bigger than the objects around it
- there is dead ground between you and the object

Objects appear further away when:

- there is low light
- the sun is in front of you
- you're lying down
- the object you're looking at is smaller than objects around it
- you're looking across a valley

To a certain extent, judging distances comes with practice. But there are a few simple techniques you can use.

DISTANCE OVER LAND

MEASURING UNITS

First off, find a unit of measurement you're comfortable with. Most people don't really know what 100 metres looks like. You might, however, be able to judge the length of a football pitch, or some other common object with which you're familiar. Three football pitches? That's approximately 300 metres.

THUMB AND ARM TECHNIQUE

Estimating the width of an object in the distance is easier than judging how far away it is. If you can do that, you can use your arm and thumb to estimate the distance. Here's how.

Let's imagine there's a building in the distance which you estimate to be 50 metres wide. To estimate how far away a point is at the same distance, hold your arm out straight with your thumb sticking up. Close one eye, and align the side of your thumb with the point you're measuring. Now, close your open eye and open your closed eye. Your thumb will appear to move. Compare the displacement with the width of your reference building. For example, if your thumb appears to have moved the width of three buildings, its relative displacement is 150 metres. Multiply this by ten to give an estimated distance: in this case, 1,500 metres.

For this technique it's good to be able to estimate the size of some common objects you might see in the distance. So: a car is about 3 to 3.5 metres long, and a mature tree trunk is about half a metre wide.

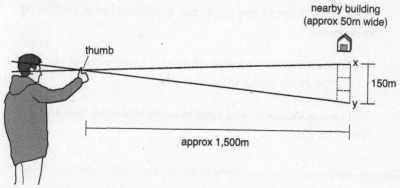

nearby building
(approx 50m wide)

thumb

x

150m

y

approx 1,500m

APPEARANCE METHOD

If you can see a person in the distance, how detailed they appear will give you an idea of how far away they are:

- 100m: all details clear
- 200m: skin colour just identifiable
- 300m: body outline clear, most other details blurred
- 400m: body outline clear, *all* other details blurred
- 500m: body appears tapered, head indistinct
- 600m: body wedge-shaped, head cannot be seen

DISTANCE OVER SEA

Estimating distances at sea involves knowing the distance to the horizon. This can be tough because how far away the horizon appears depends on your height above sea level. There are complicated calculations you can do to work this out, but I just remember that if my eyes are six feet above sea level, the horizon is about three miles away.

Once you know this, you can judge how far away an object is at sea by estimating what proportion of the distance it is to the horizon.

ESTIMATING TIME

In a survival situation, the crucial pieces of information you'll need to know are:

1. How long it will take you to cover a given piece of terrain.
2. How much sunlight you have left.

In these situations, you need to know Naismith's Rule and the Rule of Hands.

NAISMITH'S RULE

This rule states that you should allow one hour for every 5 kilometres going forward, and an extra half-hour for every 600 metres of ascent. So: 10 kilometres with an ascent of 600 metres should take about two and a half hours.

Clearly, this is very much a rule of thumb. People who are experienced in the wild will learn to adjust it according to their own fitness levels and pace. If you're fitter and/or fresher, you'll be quicker; if you're unfit and/or exhausted and dehydrated, you'll be slower. Groups tend to travel more slowly than individuals because a group has a limiting factor: the speed of its slowest member. But Naismith's Rule is enough to give you a rough idea of how long it will take to cover a given section of a map, or a stretch of terrain whose distance you have estimated.

During SAS selection, we worked on moving 4km an hour over the mountains, day and night, rain, sleet or snow, while carrying up to 75lb of gear. It was a push!

RULE OF HANDS

You can estimate how long it will be before darkness falls by measuring how many fingers there are between the bottom of the sun and the horizon (don't include your thumb). Each finger measures approximately fifteen minutes. This means, neatly, that a hand is about an hour.

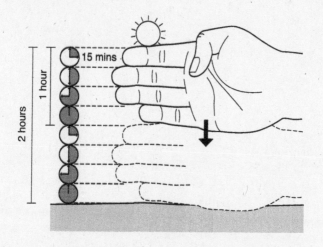

K.I.S.S.
(KEEP IT SIMPLE, STUPID!)

Distances can be very deceiving. Even experienced outdoorsmen can make estimating errors.

Use your arm and thumb to judge distances over land.

Know the distance to the horizon across water for your height.

Tailor Naismith's Rule individually before you find yourself in a survival situation.

HOW TO

COMMUNICATE

IN A

SURVIVAL

SITUATION

Sometimes, the only way you're going to survive is if you receive help from the outside world. It's essential that you know how to alert people to your predicament and, if and when help does come, that you can communicate effectively with them. Here's how.

FIRE SIGNALS

If you know that people are out looking for you, three small fires, set in a triangular formation, are an internationally recognized distress signal.

It's entirely possible, however, that you'll be trying to alert people who *don't* know you're in trouble, or that you're under canopy. In which case your best bet is to use smoke as a distress signal.

Forget all that stuff about Native American smoke signals. The trick is to make a fire that produces as much smoke as possible. If you're in the jungle, you need to make a plume thick and powerful enough to penetrate the canopy. In the section on fire (see pages 33–9), I gave the advice that small fires are best. A signalling fire is the exception to the rule. It needs to be as big as you can safely make it.

Collect enough fuel gradually to build up an enormous blaze. When it's really going strong, smother it with loads of green matter. Now make sure that air gets thrust in from the bottom. Really go for it. Waving a jacket at the base of the fire will do. Even better, use a big sheet of metal or palm fronds. The aim is to produce a big black plume of smoke that can be seen for miles around.

Smoke is only useful during the day. At night, signal fires are better. If you can find a lone tree, making a fire underneath it should ignite the whole tree and make an enormous blaze. But only do this if there's no risk of the fire spreading, which will put your life in far greater danger (see pages 333–8 for how to escape a forest fire).

MORSE CODE

You don't really need to learn Morse code like an expert, but don't underestimate how useful the SOS signal is: three short taps, three long taps, three short taps; or three short flashes of light, three long ones, then three short ones. You can use it to signal to aircraft, or to alert people to your presence if you're trapped in a collapsed building, caught in an avalanche or stuck in a cave system. SOS has probably saved more lives than all the other distress signals put together.

WHISTLE BLASTS

The sound of a good whistle can reach for miles around, and will exhaust you less than trying to shout. The international distress signal for whistles is six blasts repeated every minute, and three blasts as a reply. Combine it with the SOS Morse signal.

MIRROR SIGNALS

You can use a hand mirror to reflect the sun and transmit the SOS Morse signal. Mirror flashes can be seen by air crews 50 miles away. If you don't have a mirror, improvise with any shiny surface. A tin can, a piece of foil, the blade of your knife or even a belt buckle will reflect sunlight.

To aim your signal, hold the reflective surface in one hand near your face and make a V-sign with your other hand, arm outstretched. Position the V so that the aircraft whose attention you are trying to grab is between your fingers. Now angle the reflective surface so the reflection of the sun hits the V (and therefore the aircraft).

SIGNALLING TO AIR CREWS

There are standard emergency signals that are understood by air crews around the world, and there are two methods you can use.

Ground-to-air signals should be marked out on the ground. Use whatever you can – rocks, logs, clothes, or you can make grooves in the earth. They should be large – at least 3 metres wide by 10 metres long, with 3 metres between each one. These are the most useful ones:

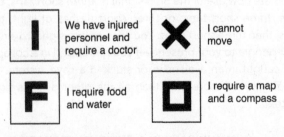

I — We have injured personnel and require a doctor

X — I cannot move

F — I require food and water

☐ — I require a map and a compass

Medical supplies required

 We are moving in this direction

 Please indicate in which direction we should move

 I think this is a safe landing zone

Or you can make distress symbols with your body. These are the ones you want:

Pick us up

We need mechanical help

All is well

We need medical assistance

Land here

Do not attempt to land here

AIRCRAFT ACKNOWLEDGEMENTS

If a fixed-wing aircraft has seen your signal, the pilot may make the following responses:

- wings rocking from side to side (daytime) or green flashes (night-time): message received and understood
- flying in a clockwise circle (daytime) or red flashes (night-time): message received but not understood

HELICOPTER MARSHALLING

If there's a helicopter coming in to rescue you, you'll increase your chances of getting out alive if you know how to marshall it safely to the ground.

First off, you need to make sure a helicopter has a safe landing zone (LZ). This means selecting a wide open flat space. Make sure any obstacles are moved out of the way, that the ground is as level as you can make it and that there is ample space for the chopper to get to the ground without any risk to its tail rotors.

Here are the basic helicopter marshalling signals. Issue these instructions, or stand with your arms aloft, and always have your back facing towards the wind. Helicopters need to land into the wind.

Hover   Move up   Move down

Land   Come and land
over here

Then, as the chopper comes in to land, make sure all hats or loose garments are off and secured, stay low, and always approach the chopper from the front. Imagine that the nose of the aircraft is at twelve o'clock – you need to approach between ten o'clock and two o'clock. Above all, stay away from the tail rotor blade and be aware of the height of the spinning main rotor blades if approaching from undulating ground.

K.I.S.S.
(KEEP IT SIMPLE, STUPID!)

Thick plumes of smoke can be created from large fires and plenty of green matter.

Everybody understands ... / – – – / ...

Ground-to-air and body signals are worth learning.

Knowing how to marshall a helicopter increases your chances of getting out alive.

HOW TO

IDENTIFY

POISONOUS

PLANTS

In a survival situation, plants fall into two categories: 'edibles' and 'deadibles'. Choose the right plant, you'll get much-needed nutrition. Choose the wrong one and you'll end up severely incapacitated, or dead. Don't underestimate just how dangerous certain plants can be.

Sometimes, though, foraged plants are all you have to eat. To stack the deck in your favour, you need to know the six-stage edibility test. This is a way of gradually establishing whether a given plant is likely to cause you harm.

THE SIX-STAGE EDIBILITY TEST

If you're in a group, you should only let one person do this – and he or she should stop experimenting with whatever plant matter it is at the first sign of anything wrong.

Take a small amount of the plant you're considering eating, and then:

1. Look at it. Does it look discoloured? Does it look old? Manky? Is it brightly coloured? (That's often a plant's way of saying 'don't eat me'.) If it's any of these, look elsewhere for your supper.
2. Smell it. Does it smell rancid or dirty? Does it smell of almonds or pears? That's often a warning sign in the natural world as it indicates the presence of hydrocyanic acid. Avoid anything with that odour (unless it's an almond or a pear, of course).
3. Do a skin-contact test. Crush the part of the plant you want to eat and rub the juice on a sensitive part of your skin – your palm, your wrist, the crook of your arm or the soft flesh by your armpit. Leave it for a couple of hours and see if you get any reaction. If your skin gets red and sore, think about what the plant might do to your insides.
4. Dab a bit of the plant on the inside of your lips. You're on the look-out for tingling or swelling. If you get any, rinse your mouth out with fresh water. (Bear in mind that the plant might taste disgusting, but that's different to being poisonous.)
5. Chew a bit. Get all the juice out of it so that it dissipates into your mouth, then spit it out. Wait to see if you have any tingling, swelling or other adverse reactions.
6. Finally, eat a small amount. Wait four to six hours and see what happens. Don't drink any water as this might dilute the poisonous effect of the plant and give you a false negative response. Don't eat any other food either, because if you do get sick you won't be in a position to know which food had the bad effect.

If you get through all six stages of the edibility test, you can be reasonably certain that the plant is safe to eat in small quantities.

WHAT TO DO IF YOU'RE POISONED

First, be very aware of the symptoms of poisoning:

- sickness
- palpitations
- seizures
- dizziness
- breathing difficulties
- drowsiness
- loss of consciousness

If you suspect that you or one of your team have been poisoned, your first aim should be to get proper medical care. If you can save a sample of the suspected poison, that will help the first responders.

If there's no hope of medical help, get yourself or your patient into the recovery position. The head should face downward to avoid choking on vomit.

Deal with the symptoms in the following way:

1. Drink plenty of water, warm if possible.
2. Try to make yourself or your patient vomit by putting your fingers down the back of the throat. Swallowing a little cold charcoal from a campfire can also act as an emetic, and may stop your blood absorbing certain poisons. Crush the charcoal and mix with fresh water. Likewise, white ash mixed with fresh water can ease stomach cramps and is sometimes an antidote to hydrocyanic acid poisoning.

EDIBLES AND DEADIBLES

It's worth knowing a few plants that can save your life – or, potentially, end it.

EDIBLES
Stinging nettles
These are nutritious, and taste good when cooked, during which process the sting disappears. But you can eat them raw too. A nettle won't sting if you pick it hard, squeeze it hard, put it straight in your mouth and eat it quickly.

Dandelions
Abundant and easily recognizable. Older ones get very bitter so are better boiled.

Acorns
They need to be soaked in water for a few hours to leach out the tannin. Sometimes they have a little grub inside. Don't waste it – the grub is edible too.

DEADIBLES
Deadly nightshade
Long, green leaves; purple, bell-shaped flowers; shiny black berries. All parts are poisonous. Symptoms include convulsions, hallucinations, coma and death.

Oleander
Evergreen with pink, white, yellow and red almond-smelling flowers. Very common in warm climates. A single leaf can kill you. The toxins attack your nervous system, your cardiovascular system and your digestive system.

Yew trees

Long, narrow, pointed leaves with
bright red berry-like fruits. The
leaves can cause heart failure
and death. Other symptoms
are muscle tremors, breathing
problems and convulsions.

MUSHROOMS

There's a saying: you can be an old mushroom hunter, or a bold
mushroom hunter, but you can't be an old, bold mushroom hunter.
If you come across mushrooms in a survival situation, you should
only eat them if you are *100 per cent* certain that you can identify
them and know they are safe to eat. Many fungi are deadly and
there's no cure. Don't risk it.

K.I.S.S.
(KEEP IT SIMPLE, STUPID!)

Know the six-stage edibility test.

Suspected poisoning? Lots of water and induce vomiting.

Know some edibles and deadibles.

Don't take a chance with mushrooms.

HOW TO
CATCH
A FISH

Fish are one of the best survival foods. They taste good, unlike a lot of the stuff you might end up having to put down your throat in order to keep alive, and they have the right balance of proteins, fats and vitamins to keep you going.

But the reality is that they are hard to catch, as anyone who's ever spent hours on a river bank with a rod and line will tell you. What's more, in a survival situation you're unlikely to have all the fancy floats, hooks and bait that fishermen use to make their lives easier. So I'm not here to tell you how to have a pleasant afternoon's fishing. I'm going to show you some methods that will massively increase your chances of catching some potentially life-saving food.

A word of warning: some of these methods, particularly night lines and gill netting, are illegal in certain parts of the world. These fishing methods are not for recreation, they're for getting your hands on much-needed food when your life depends on it. Some require basic gear that you might have with you. Others don't.

NIGHT LINES

To make a night line, attach several baited hooks to a long piece of fishing line, as shown. You'll need to attach a heavy weight to one end of the line and anchor the post to a fixed point on the river bank. Leave it there for several hours (it can be overnight, but night lines can of course be used during the day).

If you don't have fishing hooks, you can improvise with thorns, or cut and sharpen a splinter of wood or animal bone.

You can weight the end of the line with a stone and put hooks all the way up the line to attempt to catch bottom, mid and surface feeders.

To bait the hooks, you can dig up worms, use insects, or attach scraps of food – meat is particularly effective.

Night lines are smart because you can set multiple lines and they work for you while you are either sleeping or doing some other task.

GILL NETTING

If you have a net, this is the most effective way of using it. It involves setting up a 'wall' of netting which the fish swim into and get tangled. You'll need to weight the bottom of the net with stones, and set floats along the top. This is a very effective way of catching fish, so you should be sure not to leave it up too long or catch more than you need.

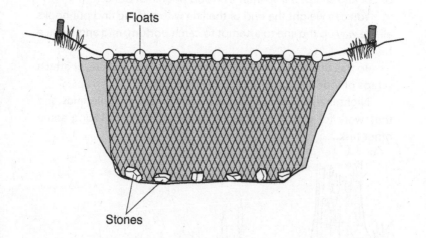

Floats

Stones

FISH TRAPS

It's possible to make an intricate basket trap from long, flexible strips of sapling. But it's a time-consuming job. You can achieve the same result with an empty plastic bottle. Cut around the top of the bottle with a sharp knife. Now ease the top of the bottle, nozzle down, back into the bottom. You've now made a trap that small fish can easily swim into, but which they can't easily escape. Pierce a few holes in the trap and, if you have it, drop some bait into it. A few stones will weight it to the river bed, and you can stop it from being swept away by attaching it to a bit of fishing line or a length of vine. Place it in shallow water, with the opening facing the direction of the current. If you can, set several of these traps. You'll only catch small fish, but in a survival situation every mouthful of food is valuable. And lots of small fish make a big meal.

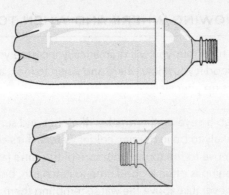

FIXED TRAPS

If you've no line, net or bottles with you, fixed traps are your best bet. These work on a similar principle to the bottle trap – you design a shape that fish can easily enter, but can't leave. You can build such traps using rocks or sticks. Just make sure that there are as few gaps in the wall as possible, and that the funnel is facing the direction of the current.

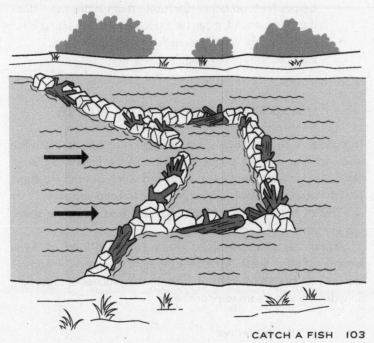

KNOWING WHERE AND WHEN TO FISH

Fishing is hard, but you will dramatically increase your chances of catching food if you know where and when to look, and have a few extra tricks up your sleeve.

- Keep an eye on the weather. If it's hot and sunny, fish are likely to be in deeper, colder waters. If it's cold, they'll move to the top to get more of the sun's rays.
- Evening is often a good time to catch fish, because insects hover just above the water, tempting them to the top.
- In general, fish are more likely to be found in the shelter of large rocks.
- Fish are attracted to areas of water that have overhanging trees, because berries and fruit – i.e. food sources – fall into the water. So, these are good areas to fish.
- Scattering bait – worms, insects, berries, scraps of food – on to the surface of the water can attract fish to that area and work them into a feeding frenzy. Be judicious – it may be that you're better off eating the insects or berries for food rather than using them in this way – but if bait is plentiful, it can be a good way of attracting fish.
- If you've caught and gutted a fish, don't discard the innards. They make great bait.

USING FISH TO GET FRESH WATER

This works even for saltwater fish (for advice on how to catch fish while adrift at sea, see pages 189–90). Cut the flesh into pieces, place them in a piece of clean cloth and wring out the moisture. That precious moisture is drinkable, and can keep you alive.

There's water in the eyeballs too, and fluid along the spine. Cut the fish in half to get to it.

Remember, you need fluids before you need food, so if you are short of other fluids, don't drink blood: it is protein rich and will use up your body's reserves of water in the process of digestion, dehydrating you. Learn to prioritize.

K.I.S.S.
(KEEP IT SIMPLE, STUPID!)

Recreational fishing is sportsmanlike. Survival fishing is about getting food, fast.

Night lines. Gill netting. Fish traps.

Know your environment, and how it affects the fish.

Fish smart and be patient.

HOW TO

TRACK,

AND TRAP,

ANIMALS

FOR FOOD

You can spend a lifetime learning how to track animals, and people do. It is a wonderful art. But in a survival situation you need to forget any image you might have of tracking reindeer for days on end across glaciers. You need to know how to find food quickly, and trap it with the minimum expenditure of energy.

With this in mind, remember that unless you're an expert, hunting is hard. Finding meat in the wild can take a long time and be energy-intensive. Gathering edible plants or even insects is usually an easier way of getting the fuel your body needs. But the smart survivor plans for both, and if you can hunt and kill an animal, you'll be getting a good source of energy-rich food which will help in your fight to stay alive.

The flesh of all furred mammals, and all birds, is edible.

WHAT ARE YOU TRACKING?

I could give you a list of a hundred different animal footprints to learn, but you are unlikely to remember them in a hurry when it matters. The ground needs to be of just the right softness to take the print, and it takes a lot of experience before you can start identifying them properly.

To a certain degree, however, you don't really *need* to be an expert in what animals you're tracking. Obviously you are not planning to start tracking a lion, but in general if you can find evidence of any wild animals at all, you're in with a better chance of getting food. So if you see animal prints, that's great; but a better strategy is to look for other marks that they leave on the landscape.

ANIMAL RUNS AND TRAILS

You need to know that, in general, animals don't travel too far. Their principal routes are going to be from their burrow or den to their food or water source, and back again. This means that they'll be regularly following the same pathways. As they follow these routes, they'll leave a path, with the vegetation bent in the direction in which the animal travelled.

There's a difference between a run and a trail. A run is used by a single animal, which means it's smaller and narrower. A trail is used by a number of different animals, so it's wider. You'll have a better chance of tracking animals by a trail, simply because there's likely to be more of them.

DENS

You're probably used to seeing animal holes if you've spent any time in the wild or in the countryside. They're not hard to spot. The trouble is, they're often long abandoned. Use your common sense.

- If you see a spider's web over the hole, or if it's covered with vegetation, chances are nothing's come in and out of it for a while.

- If you see fresh faeces nearby, or teeth marks on the surrounding vegetation, it's likely to be an active den.

Some animals, like deer and hares, don't live in dens. But they've got to sleep somewhere, and you can often find the impression of where they last laid their body in the foliage.

POO

Experienced trackers can identify animals by their poo. In a survival situation we examine poo not to work out *what* we're hunting, but to work out *how recently* it passed by. Fresh equals recent.

THINK LIKE AN ANIMAL

Wild animals want to survive too. This means they are incredibly sensitive to sound, smell and movement. So if you're going to track them, remember:

1. To an animal, you stink. Approach downwind to stop your scent being carried.
2. Your footsteps can be heard from a great distance. Move lightly.
3. If you're waiting for an animal to appear, keep completely still. The smallest movement will scare them away.

MAKING THE KILL

Knowing where animals are likely to be is one thing. Catching them is quite another. If they see you, smell you or hear you, they'll be off. So you need to think a bit smarter. If your presence is a deterrent, you need to catch the animals when you're not in the vicinity. This means having a few simple traps and snares up your sleeve.

There are lots of different types of snares and traps. I like to think of them as falling into four categories: strangle, dangle, tangle and mangle. Be warned: these are neither pretty nor humane. Survival doesn't work that way. Only use these snares if it's a matter of life or death.

And the more traps you set, the better your chance of making a kill.

A SIMPLE SNARE (STRANGLE)

This snare is a loop that will tighten as your animal steps into it. The more the animal kicks and struggles, the tighter the snare will become. The best material for this snare is a piece of wire: you can easily twist it to make a loop as shown in the illustration below, it resists becoming looser, and is harder for the animal to gnaw its way through.

You can hang a snare like this over the mouth of an active animal hole, or you can suspend it on a couple of twigs as shown and place it on an animal run.

Remember always to secure a snare firmly to something that won't get ripped away. A struggling animal generates a lot of force.

A SIMPLE SPRING SNARE (DANGLE)

This is an adaptation of the simple snare, and is suitable for an area where there are lots of strong, springy saplings close to an animal run or trail. Use your knife to make a notch as shown that will dislodge easily when the snare is disturbed. The advantage of this kind of snare is that it raises your prey off the ground where it is less likely to escape or be taken by other   predators.

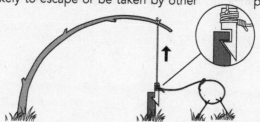

A SIMPLE TREE SNARE (TANGLE)

This is good for squirrels and tree-climbing animals. Look for a tree that leans at an angle. If you can see scratch marks on it, it's a sign that animals regularly climb it. Now, find a pole that you can lean against the tree. The animal will be more likely to take this less steep route into the tree. Cover it with small simple snares. The animal will get tangled in these – and again, the more you use, the higher the likelihood of tangling them.

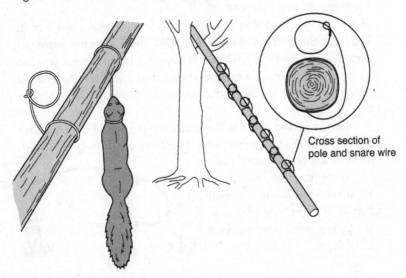

Cross section of pole and snare wire

A PITFALL TRAP (MANGLE)

This is quite an energy-intensive trap, but it has the benefit of not requiring anything in the way of materials. You simply dig a deep hole in the middle of an animal run or trail and cover it with a network of branches and moss. If you have some bait, rest it on the top. And to make the trap more deadly, you can place sharpened sticks in the bottom, pointing upward, to spear the animal as it falls in.

K.I.S.S.

(KEEP IT SIMPLE, STUPID!)

Forget about tracking long distances. Animals go from their dens to their feeding and watering places and back again.

Learn to recognize runs and trails.

You stink. You're noisy. You're clumsy. So use your superior brain.

Strangle. Dangle. Tangle. Mangle.

HOW TO
PRESERVE
FOOD

In a survival situation, if you manage to catch meat or fish, you've found yourself a precious resource. It's important to ration this resource carefully, because food might be hard to come by in the days ahead. When you're keeping yourself alive, you cannot waste fuel.

However, these foodstuffs spoil quickly. Once a piece of meat or fish has turned rancid, you'll do yourself more harm than good eating it. But if you know how to preserve food, you can keep it going for months. Here are the best ways to do it.

AIR-DRYING

This is a good way of preserving meat in sunny environments. Small animals, up to the size of a squirrel, and small fish can be dried whole in the following way:

1. Bleed, gut and skin the animal. Gut fish (by cutting along the belly and scooping out the insides) and remove their heads and tails.
2. Hang the carcass in a sunny place where there is decent airflow.
3. When the outside of the meat feels completely dry, break up the carcass with a rock so that the insides – the bones and the marrow – are exposed. They won't have dried through yet.
4. Hang the carcass again, and leave it to dry for a second time.

If you have lean flesh:

1. Cut it into strips about 3cm long and half a centimetre wide.
2. Trim off as much fat as you can.
3. Make a drying rack. There are lots of complicated ways of doing this, but two branched sticks stuck in the ground with a pole between them will do the job.
4. Hang the strips so that they are not touching each other.
5. Leave in the sun till the meat is crisp enough to snap – about twenty-four hours should do it.

SMOKING

If it's not dry and sunny, air-drying won't be possible. But you can still preserve strips of meat – and small fish – by smoking them. This will both dry the food out and create an anti-microbial layer on the outside. Here's the easiest way:

1. Dig a pit, and light a fire in the bottom (see pages 33–9).
2. When the fire has burned down to embers, add green twigs. If they seem very dry, soak them first.
3. Suspend your meat above the fire as for air-drying.
4. Don't let the fire get too hot – you're smoking, not cooking.
5. Keep the smoke going by gradually feeding the fire with more green matter until the meat cracks when you bend it – again, it can take up to twenty-four hours.

SALTING

This is one of the oldest ways of preserving food. In times gone by, the Royal Navy diet centred around salted meat that could last on long voyages. It's not particularly practical in a survival situation, but if you're by the sea and can boil away sea water until you're left with a residue of salt, you can rub this into your meat or fish before air-drying or smoking it. This speeds up the process, and can make the finished product taste better. Wipe off any excess salt – or soak the meat/fish in clean water if you have enough – before eating it.

FREEZING

In Arctic environments, nature has provided you with a ready-made food preservation system: snow and ice. You need to skin and gut your meat in the usual way, then cut it into small pieces so that it freezes quickly, before digging a hole and packing it in snow and ice. Once it's thawed, don't refreeze it.

PEMMICAN

Pemmican tastes horrible, and is hardly a healthy way to live for long. But for anyone venturing out into the wild, this mixture of preserved meat, fruit and fat is the way ahead. This is the stuff that kept the great Antarctic explorers alive for months. Native Americans kept it for years.

To make it, you need equal quantities of dried meat and suet, or rendered animal fat. Grind the meat down to a powder, then melt the fat and mix into the meat powder. If you have any pieces of dried fruit, add those too. You'll get a substance the same consistency as sausage meat.

Pemmican can be bitten into like an apple, or dissolved in boiling water to make a soup. Don't underestimate what an efficient survival food this is. During the Second Boer War, soldiers were expected to march for thirty-six hours on 4oz of pemmican plus 4oz of chocolate or sugar. They kept this in a little iron tin – hence 'iron rations'.

K.I.S.S.
(KEEP IT SIMPLE, STUPID!)

Food is precious. Don't gorge, preserve.

Sun. Smoke. Salt.

In the Arctic, use your natural freezer.

Make like an Antarctic explorer and learn to love pemmican.

HOW TO

HANDLE

YOURSELF

IN A

FIGHT

Fights aren't like they are in the movies – well choreographed and glamorous. People don't receive a series of uppercuts and then stagger away with little more than a woozy head and a bleeding lip. A real fight is short, ugly and incredibly violent. Be under no illusion that:

- On average it will last no longer than three seconds.
- It can result in life-altering, systemic damage to a human being.
- A single punch can either put you in a coma or kill you.
- Almost every fight will end up with you needing immediate medical attention, even if you are the winner.

So, rule number one: the best way to avoid being hurt is to avoid getting involved in the first place. Don't lower yourself to someone's drunken, aggressive level. Walk away. Or run if you have to. Remove your ego and live to see another day.

WARNING AND DANGER SIGNS

Police officers and anyone involved in conflict resolution are taught to recognize *warning signs* that someone is getting angry, and *danger signs* that they might be on the point of attacking.

Warning signs include:

- face going red
- prolonged eye contact
- raised voice
- gritted teeth
- faster breathing
- standing up tall to look bigger

Danger signs include:

- colour draining from face
- clenching and unclenching of fists
- head dropping to protect the throat
- hands rising above the waist
- shoulders tensing
- stance moving aggressively from square to sideways
- staring at intended target

Understand that these danger signs mean *the fight has already started.*

WHAT TO DO WHEN YOU SEE THE DANGER SIGNS

If someone demands your wallet, chances are that all they're after is the money. Hand it over. Live to see another day.

On the other hand, if someone is shouting abuse, prodding you or spitting aggressive language at you, your companions or your family, you need to understand that you are *already* in conflict. That person is out to hurt you and wants to inflict violence.

If you're prepared, and you remember that abusive words and danger-sign body language mean conflict is imminent, then your next move is clear: you must do whatever it takes to stay alive and unharmed. That means making a pre-emptive strike. You must enter the fray first, and win.

The law says that you can use force to protect yourself, protect others and protect property, so long as it's reasonable and necessary. As long as it fulfils these criteria, you can strike first.

ADRENAL FREEZE

Doing this isn't always as easy as it sounds. You might have experienced or witnessed a phenomenon called adrenal freeze – that sensation when a person is shocked into inaction. Everyone is highly susceptible to this if not trained: energy drains from our muscles and our mind seems to have no control over our body. Adrenal freeze is the physical manifestation of fear, and most people get hurt in fights because of it.

The best way to overcome it is with positive action. That means being confident in your own abilities in a violent situation. It means having a strategy. It means knowing *how* to fight.

IMPACT FACTORS

You, your opponent and the environment. These are the impact factors that determine how a fight is likely to unfold.

For example, your opponent is a foot taller than you, and clearly stronger. He or she might be under the influence of alcohol or drugs. You're wearing glasses and have a twisted ankle. Your opponent appears wild and aggressive.

Time to think smart.

This is where your environment comes in. Maybe there's an incidental weapon to hand – a chair, a stick, a bunch of keys. Are you inside or outside? Is there an escape route? Can you get an advantage? A heightened awareness of these impact factors will help you decide: fight or flight?

KRAV MAGA

Krav Maga is a self-defence system utilized and honed by members of the Israeli secret service and military. (Trust me: those guys know how to handle themselves.) Krav Maga is.

- highly practical
- focuses on real-world situations
- teaches you to avoid violence wherever possible
- gives a full complement of brutal, often pre-emptive, counter-attacking tactics

I want to give you the low-down on a few Krav Maga basics. There's a whole lot further you can go with this, but these few simple tools should massively increase your chances in a violent situation.

SAFE ZONE
Krav Maga teaches you never to allow an aggressor through your safe zone. Keep your hands out in front of you to hold an aggressor at bay. *Do not let them through.* If they enter that space, you push them away with a loud, dominant command: 'Back away!' If they attempt to enter the space again, you strike.

SOFT BODY PARTS AND PRESSURE POINTS
We're not trying to be chivalrous here. Your opponent won't be fighting by some code of honour, so neither should you. Street fights are not like that. One thing matters and that is winning. Successfully neutralizing an attack means going for your opponent's weakest, most vulnerable points. We're talking eyes, throat and groin. Clawing at their eyeballs, punching their throat and kneeing them in the nuts may not sound very sportsmanlike, but that ain't the point. Your aim is to win – whatever it takes.

Eyes

2. Throat

Throat

Groin

HOW TO PUNCH

Strike with a closed fist, with your thumb on the outside. Many martial arts suggest multiple types of strikes, from open hand to palm strikes, but the truth is that in every fight you need to assume your life is on the line. In which case, you must use the most reliable measures, even if the consequences are some broken knuckles. Under pressure and in the moment, you need to know that if you are relying on a strike, it will work and you won't be susceptible to an adrenal freeze response because the strike is too complicated to remember. This might sound unlikely, but it isn't. In the heat of battle – which, as I have said, generally lasts a matter of seconds – you want to rely on the most devastating, effective and simple strike. The closed-fist punch is that.

Always put your full body weight into it as you make it, to give you momentum and power.

Aim for the throat, but an accurate, explosive blow delivered correctly to any area of the head will incapacitate 99 per cent of opponents. The throat strike will stop them breathing. Any other head strike will often result in a backwards fall and concussion at least.

Always remember to keep moving positively forward, striking repeatedly, until the opponent goes down or the threat is neutralized.

Be aware of multiple opponents. As soon as one threat is down, turn fast and assess for further threats. Adrenalin can result in a tunnel vision that makes you unaware of other attackers behind or even to the side of you in a fight.

Rely on the big motor neurone skills that you know will still work when adrenalized, such as moving your whole body around to scan for other opponents, rather than just moving your head.

DEFENDING A ONE-HANDED THROAT HOLD AGAINST A WALL

Grabbing someone's throat and pushing them up to a wall is a very common attacking move. Krav Maga gives us an effective counter-attack. Tuck your chin towards your chest – this weakens your opponent's grip. Open your palms. With your weaker hand, grab your opponent's wrist near the thumb and force it down towards the ground. At the same time, drive a punch into your opponent's throat. You can follow this up with a knee to the groin.

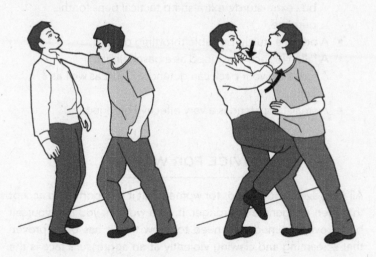

DEFENDING A CHOKE FROM BEHIND

Again, this is a very common attacking move, and a dangerous one because it cuts off your air supply. To counter-attack, raise one arm so that your elbow is at the level of your ear. Step the corresponding leg behind the other and twist your body in the direction of your raised arm (so, if you've raised your right arm, you need to twist clockwise). Do this as aggressively as possible, dropping down at the same time so you are putting as much weight as you can against your opponent's wrists. This will remove you from the choke, at which point you should either counter-attack with a series of aggressive strikes, or get out of there.

INCIDENTAL WEAPONS

Krav Maga reminds us that anything can be a weapon, if we choose to use it as one. You just have to think smart and improvise. For example:

- A house key could save your life if you use it to stab at soft body parts like the eyes.
- Same goes for a pen – sharp, easily concealed. You can buy extra-sturdy, extra-sharp tactical pens for this purpose.
- A belt is a sturdy, portable throttling device.
- A fallen branch is as good as a baseball bat.
- A fork from your pack can defend you just as well as a knife.
- A small lampstand is a very effective bludgeon.

ADVICE FOR WOMEN

All the advice above holds for women, but it is important to accept that men are generally stronger. If, as a woman, you find yourself being overwhelmed, you need to know that it has been proven that screaming and clawing violently at an aggressor's face is the best way to repel an attack. Clawing is instinctive, which will help you overcome any adrenal freeze. For the reasons I've outlined above, you should never acquiesce in the hope that you can calm your aggressor down. It doesn't work like that. Meet force with an overwhelming counter-force. Make them think twice. Make them understand that you are *not* an easy target.

Do not think you can talk aggressors down. If you see the symptoms listed at the start of this section, no amount of talking is going to work.

You can, however, use language to confuse any aggressor before a strike. I know one doorman who, before a fight, will always ask the aggressor: 'Is your mother called Wendy?' This almost always makes them stop and think for a split second. That is all he generally needs to get a strike in first.

Two things to remember never to agree to: relocation and being tied up. Both of these mean you lose all your power to determine the outcome of conflict. Fight for your life if someone is trying to bundle you into their car or tie you up (see the section on kidnapping, pages 353–60).

Finally: if an aggressor ever gets into the passenger side of your car, shout and scream aggressively and start striking them immediately with the side of your left fist into their head (assuming you are in a right-hand-drive car). If they get out, you can drive off at once. If they don't get out, keep striking, while reaching for your own seat belt with your right hand, so that you can escape the vehicle yourself. If you have children in the back, shout at them to get out of the car. Once they're out, you too can exit the vehicle.

Remember, it's always better to lose a car than your life.

K.I.S.S.
(KEEP IT SIMPLE, STUPID!)

Recognize when you are in conflict before
you are actually attacked.

Run away if you can.

Remember: you never win a defensive fight.
Strike first, strike hard.

Go for your opponent's weak spots: eyes, throat, groin.

Regularly practise simple strikes, accurately
and explosively delivered.

Win, whatever the cost.

HOW TO
HANDLE
FIREARMS
SAFELY

In many parts of the world in which you might find yourself, firearms are a fact of life. Many of these weapons are specifically designed to kill a human. Some aren't. But even shotguns and air pistols, designed for hunting and recreation, can be fatal. If you come across any of these weapons, you need to know how to make them safe.

The trouble is, there are so many different types of weapon, each with their own quirks and characteristics. Certain procedures are similar with all firearms, however. Know them:

1. *Avoidance.* If you don't need to handle a firearm, don't handle it. Whatever you do, don't make the mistake of thinking that what you see in the movies has any resemblance to real life. You can't just grab a weapon: you don't know what state it's in, or if it's ready to fire.

2. *Keep the weapon pointing away from you.* And away from anyone else, too. That means pointing it at the ground or up in the air. Remember that firearms can have a very long range. Just because you can't see someone in your line of fire, because of trees or other obstacles, it doesn't mean they're not there.

3. *Apply the safety catch.* Most weapons have one. It'll be a slide, a switch or a plunger. Some firearms have internal safeties, which means no switch. However, never assume that a firearm has been made safe if there isn't an obvious safety switch.

4. *Never let your finger anywhere near the trigger.* When handling a weapon, keep your finger on the outside of the trigger guard at all times.

5. *Take off the source of ammunition.* There will probably be a magazine of some description. Carefully remove it. With the magazine out, you need to unload any ammunition in the breech. Cock the weapon several times to make sure the breech is empty, then fire off the action in a safe direction.

You will, of course, need to adapt these rules according to the type of weapon. A revolver, for example, keeps its ammo in a cylinder; a shotgun breaks rather than having a magazine to remove. But if you remember the basic guidelines – keep it facing away, switch on the safety catch, remove ammunition, then unload – you'll be on the path to making sure you handle a weapon safely.

HOW TO FIRE A WEAPON SAFELY

If you find yourself in a position where you have to fire a weapon of any type, these are the military marksmanship principles I was taught in the SAS.

1. The position and hold must be firm enough to support the weapon.
2. The weapon must point naturally at the target without any undue physical effort.
3. Sight alignment and sight picture must be correct.
4. The shot must be released and followed through without undue disturbance to the position.

To that, I would add the following:

1. Avoid tensing up. People expect there to be a big kick, but that isn't really true of many modern weapons.
2. Avoid touching the trigger until you've made a conscious decision to shoot. Otherwise keep your finger outside the trigger guard.
3. The single most important factor when firing a weapon is breathing. When you breathe, your body moves. If you want to have any chance of hitting anything, you need to stop breathing. So: breathe in, exhale three-quarters of the breath, hold, then fire.
4. Don't hold your breath too long or your body will start to shake.
5. Get into a comfortable position with your aim, then close your eyes and relax for a few seconds. Re-open your eyes and see if you are still aiming on target. If not, adjust your position until your aim is naturally aligned.

6. Don't snatch at the trigger. Squeeze it, hold, then release.
7. With handguns, be aware of the recoil of the working parts. Keep your thumb closed around the stock, not pointing up in the air. Otherwise you risk losing it!

K.I.S.S.
(KEEP IT SIMPLE, STUPID!)

Don't handle a firearm if you don't have to.

Don't point it at anything you don't want to kill or destroy.

Make it safe. Remove the ammunition. Unload. In that order.

Don't tense up. Stop breathing. Squeeze.

Be aware of your thumb placement.

GREAT
ESCAPES

HOW TO

SURVIVE

GETTING

LOST

Millions of people get lost every year. Sometimes they lose their lives as a result. They find themselves completely unable to retrace their steps. They become dehydrated or hypothermic. Their critical faculties grow impaired and they become less able to make smart choices. It's a vicious cycle that can end in a corpse.

When that happens, it's almost always because they've made one simple mistake: they didn't stop as soon as they realized they were lost.

It's a natural human instinct to keep going. We don't like going backwards. We don't like retracing our steps. If we're intending to go from A to B, we feel we've failed if we turn back to A, so we risk ending up at C. We talk ourselves into thinking that we're going the right way.

Instead, we need to be rigorous about not fooling ourselves. We need to swallow our pride.

DON'T PANIC

That moment when you first realize you're lost can be scary. It can make people lose their heads. It can be disorientating and frightening. I have experienced it a few times and the initial urge to panic is strong. But panic decreases your chance of survival. Try to keep a lid on it. Know that a clear head helps you make smart choices. Have a breather and get some water down you. You're not lost yet, you're just temporarily confused!

A good way to stem the panic is to remember the acronym S.T.O.P.

- **S**top: don't make a bad situation worse by pushing blindly on and getting more lost.
- **T**hink: your brain is your best survival tool, so control it and use it to think logically.
- **O**bserve: if you have a map, look for big, obvious features that you can't mistake for anything else in order to orientate yourself. It might be a tall antenna, or a huge lake. *Don't* fool yourself into thinking that a small stream is definitely the one you want it to be when there are loads of streams in the vicinity.
- **P**lan: have a definite strategy, which will force you to think things through clearly and, crucially, keep your morale up. Nothing is more dispiriting than not knowing what you're doing or where you are going.

RETRACE YOUR STEPS

This is the most important thing you can do when you're lost. Don't keep wandering blindly into the unknown. The chances of you stumbling across the right path are minuscule. If you can retrace your steps to the last point where you were sure of your location, then job done. Do this by looking out for landmarks that you recognize, or by re-walking an existing trail, or one that you've made. Don't let your pride get in the way.

WHAT IF YOU'RE STILL LOST?

If you're still lost having retraced your steps, here's what you need to remember.

1. **Conserve your energy.** See pages 53–7 for more on this, because it's crucial that you don't exhaust yourself. If you truly are lost, you could be in this for the long haul. You'll need every ounce of energy you can muster.

2. **Keep track of time.** Trying to find your way in the dark is a bad idea. If it looks like you'll have to bed down for the night, make sure you have enough daylight to find or make shelter.

3. **If you have a rucksack, always keep it on you.** Don't put it down then go and do a recce. If you can't find it again, you'll be in a worse position than before. Remember the Commando ethos: never get separated from your kit.

4. **Head for water.** Civilizations live near rivers. If you find a water source, follow it downstream.

5. **Mark your trail.** Even when you're lost, it's a good idea to mark your path so that you can retrace your steps quickly and easily if needed. Make sure your markers are distinctive, so you can easily find them again.

6. **Look back.** So you know what your path looks like going the other way.

7. **Get up high.** If you can climb a tree or a hill, you'll get a better idea of the topography of the area than you would at ground level. Don't take unnecessary risks, however, and don't stray too far to do this without knowing that you can retrace your steps.

8. **Follow power lines.** These can stretch across vast uninhabited areas. But eventually they lead somewhere.

9. **Light a smoky fire.** If people are out looking for you, a signal fire will give them something to head for (see page 87 for the low-down).

10. **Use a whistle.** Shouting to get people's attention is exhausting and you'll soon lose your voice. The sharp

toots of a whistle carry much further than a voice and you'll use less energy (see page 88).

11. **Work out your cardinal points.** See pages 71–8 for tips on finding north, south, east and west without a compass. This might help you relocate a village you've travelled through, or some point of safety.

K.I.S.S.
(KEEP IT SIMPLE, STUPID!)

S.T.O.P.

Retrace your steps.

Conserve your energy, and keep track of time.

Head for water and follow it downstream.

HOW TO

HITCHHIKE

SAFELY

AND

EFFECTIVELY

Hitchhiking can be a fast, efficient and cheap way to travel across the world. Ninety-nine times out of a hundred it's completely safe. But it's not without its risks. People have been abducted, abused and murdered while hitchhiking. If you're going to do it, here's how to minimize the risks.

CHOOSE YOUR SPOT CAREFULLY

If you're hitching a ride, you have to put yourself by the side of a busy road. The priorities of survival apply here just as much as they do in the jungle.

A roadside is a dangerous place, and this is probably the most hazardous part of hitchhiking. If you can avoid it, by hitching rides at petrol stations for instance, do that.

If you can't, choose a spot where you don't have to stand too close to fast-moving traffic, preferably by a tree, streetlight, telegraph pole or sturdy sign, which will take the brunt of impact from any vehicle that accidentally leaves the road. Make sure there's enough room for a driver to pull over safely. Wear bright, preferably reflective, clothes to ensure that you're visible to oncoming traffic.

MAKE THE RIGHT SIGN

The standard hitchhiking gesture is a thumbs-up sign, but be careful: it's a massive insult in certain parts of the world, such as the Middle East, West Africa and South America. Arm yourself with local knowledge first.

It's much better to get a piece of cardboard and write your destination on it in big, clear letters that can be seen from a moving vehicle. You're more likely to get a ride if the driver knows where you are going, and it's more likely to be a safe ride too.

TRUST YOUR INSTINCTS
AND CHOOSE YOUR RIDE

In a hitchhiking situation, *you* need to maintain control. That means trusting your instincts. If you feel in any way uncomfortable about the car or the driver, don't get in – even if you've been standing on the roadside for hours and you're desperate for a ride. Remember: you choose the ride, the ride doesn't choose you.

Be alert for warning signs of an untrustworthy driver:

- They avoid the question when you ask them where they're travelling and why (you should always do this).
- They avoid looking you in the eye, or maintain too much eye contact.
- They seem too keen – a trustworthy driver ought to be as wary of you as you are of them.
- They seem impatient, uncomfortable or controlling.
- The car is unusually dirty or smelly, or contains empty alcohol bottles or cans.

If you decide to turn a ride down, make a polite excuse to avoid antagonizing the driver. Say that you suddenly feel unwell, or that you've just remembered you left something at your previous location. Give them the opportunity to remove themselves from the encounter with dignity.

If you decide to take the ride, explain to the driver that you are going to text their number plate to a friend. If they are trustworthy, they won't mind you doing that. (You can pretend to do this if you have no battery or signal.)

Always agree in detail where you're going to be dropped off before you accept a ride. That way the driver won't be able to become vague about it later on.

ONCE YOU'RE IN THE VEHICLE

Keep your most valuable items on your person – think passport, wallet, money and phone. If you have to get out of there in a hurry, you might have to leave your pack.

Be polite but confident. Don't acquiesce to anything you don't want to do, and don't agree to being taken off your agreed route.

Avoid discussing potentially inflammatory subjects, such as politics or religion.

Stay awake, even if you're exhausted. As soon as you fall asleep, you become an easy target.

If you feel uncomfortable and want to get out of the vehicle quickly, tell the driver that you feel unwell and you're going to vomit. Once they stop to let you out, run in the opposite direction to the one in which you've been travelling.

K.I.S.S.
(KEEP IT SIMPLE, STUPID!)

The most dangerous element of hitchhiking
is standing close to moving vehicles.

You choose the ride, the ride doesn't choose you.

Text your driver's vehicle details to a friend (or pretend to).

If you feel uncomfortable, say you're going to vomit.

HOW TO

CARRY

OUT

BASIC

MOTOR

MAINTENANCE

People mistakenly assume that their vehicle is a safe place. Statistically, it's one of the most dangerous – not only because of the risk of road traffic accidents, but also because we use them to travel far from safety, often without much thought as to what might happen if we got into trouble.

For this reason, there are certain core motor skills every driver should have. You can't rely on twenty-four-hour breakdown services when you're off grid and miles from civilization.

CHANGING A WHEEL

Kill your engine and engage the handbrake. Put the vehicle into first gear, or Park if it's an automatic transmission. If you're on a slope, it's a good idea to chock your wheels to stop the vehicle rolling. You can get special tools to do this, or improvise using a piece of wood or a rock from the side of the road.

In the back of your vehicle you should have: a spare wheel, a wheel-nut wrench (with a special adapter if your wheels have 'locking' wheel nuts) and a jack. Remove them all and lay the spare tyre on the ground next to the flat one. Lever off the wheel trim or hub cap on the flat tyre so you can get to the wheel nuts.

Your vehicle handbook will show you where the jack needs to be positioned. Don't put it just anywhere – you'll damage the vehicle and get yourself into more trouble.

Use the wrench to crack the wheel nuts loose, but don't fully remove them yet. (You may need to use the locking wheel-nut adapter for one of them.)

Now, raise the jack so the vehicle is just off the ground. You don't want it to be sky high but it needs to be high enough to allow the more inflated spare tyre to fit on.

Remove the nuts, leaving the top one till last. Place them in a secure, clear place. Now remove the wheel.

Reverse this procedure to fit the spare. Tighten the nuts by hand first, using a diagonal pattern, then tighten them to a pinch using the wrench. Lower the jack so the wheel sits on the ground. Finish by fully tightening the nuts using the wrench, again in a diagonal pattern.

Stow the flat tyre in the boot. Be aware that the spare may only be good for temporary use and at a restricted maximum speed – get it replaced as soon as you can.

Remember: never put any part of your body under a car that's only supported by a jack.

OIL

You should always check your oil before setting out on a long journey, especially if it's likely to take you off grid. If your oil pressure or level light comes on while you're driving, you need to stop the vehicle immediately: oil starvation can cause catastrophic engine damage, and such damage can occur in under a minute. If your engine seizes up while you're driving, you're at risk of a potentially fatal MVC (motor vehicle collision).

OIL CHECK

Oil should always be checked with the vehicle on level ground. Allow at least ten minutes after stopping a hot engine, but preferably check it when cold.

Open the vehicle's bonnet and locate the dipstick – it normally has a round or T-shaped handle. Check the vehicle's manual if you can't locate it.

Remove the dipstick and wipe it clean. Re-insert it, then remove it again. Look for the two marks at the end of the dipstick. They indicate the minimum and maximum oil levels. If you're under the minimum level, you need to add more oil.

ADDING OIL

First, you need the right oil. Your vehicle manual will tell you what it is. It might recommend a particular brand, but ignore that: the information you need is the grade (for example 5w40) and the ACEA specification (for example ACEA A1/B1).

Locate the oil filler cap. It normally has a representation of an oil can on it, but check the manual if you're not sure. *Don't* get this wrong: if you pour oil into the coolant reservoir by mistake you'll potentially cause irreversible damage to the engine.

Add the oil in small quantities, checking the level on the dipstick each time you add some, until you get to the maximum mark. On most vehicles (but not all) the difference between the minimum and maximum marks is about a litre. Don't attempt to fill to the top of the filler cap – you'll cause hydraulic lock and massive engine damage.

Be clear on the difference between the symbols for windscreen fluid, coolant and oil. And *never* attempt to remove the radiator cap when the engine is hot or you will get severely scalded. Also, tempting though it might be, never pour cold water or coolant into a hot or overheating engine. The resulting thermal shock can crack your engine.

OIL CHANGE

From time to time the oil will need to be drained from the engine and new oil added – it's routine maintenance. If you find yourself in a situation where your vehicle's oil needs changing and there's no help around, here's how to do it.

1. Run your vehicle for at least ten minutes. This heats up the oil, lowering its viscosity and allowing it to drain more quickly and completely from all parts of the engine.
2. Check your vehicle manual for the location of the oil drain plug (usually located on the bottom of the engine's oil sump).
3. Jack up the car in the appropriate location (see above) and place a container under the drain plug.
4. Wearing gloves, undo the drain plug and allow the oil to drain completely. The oil will be hot as it flows into the container.
5. Once it has drained, wipe the underside of the sump and replace the drain plug.

If possible, you should also replace the oil filter. This can be a paper element type located in a plastic housing under the bonnet, or a metal canister type normally located near the lower half of the engine and accessible from underneath (check your manual). Move the drained-oil container so that it's underneath the filter and catches the small amount of oil trapped in the filter canister itself. Unscrew the filter and allow any oil to drain fully. Smear some fresh engine oil on to the rubber seal of the new oil filter and tighten by hand to a good pinch.

JUMP-STARTING

If your vehicle's battery has died, you can jump-start it from another vehicle using a set of jump leads. To do this:

1. Place both vehicles in neutral or Park and shut down the ignition.
2. Attach one of the red jump lead clips to the positive terminal of the working battery (it will have a + sign or the letters 'POS').
3. Attach the other red clip to the positive terminal of the dead battery.
4. Attach a black clip to the negative terminal of the working battery (it will have a − sign or the letters 'NEG').
5. Attach the other black clip to a good earth, such as an exposed bolt-head or designated earthing point (look for a label with an earth symbol, or an exposed copper bolt-head). There may be a little sparking so always use the designated earthing point or a suitable earth away from obvious sources of ignition (anything near the fuel system for example).

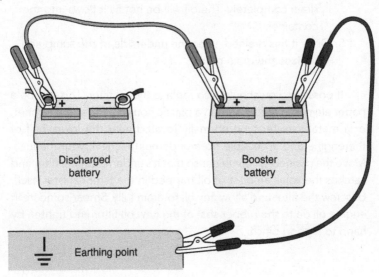

Discharged battery

Booster battery

Earthing point

- Start the ignition of the vehicle with the good battery. Let it run for a few minutes.
- Try to start the vehicle with the dead battery. If it starts, you can remove the cables in reverse order.
- Don't switch off your engine straight away. Drive your vehicle around for at least fifteen minutes to put some charge back in the battery.

K.I.S.S.
(KEEP IT SIMPLE, STUPID!)

Don't be too proud to consult your vehicle manual.

Only jack your vehicle at the dedicated jacking points.

Know the difference in symbols between coolant, windscreen fluid and oil.

Never remove a radiator cap when the engine is hot.

Know the order in which to attach jump leads.

HOW TO

DRIVE

OFF

ROAD

I learned a lot about driving when I joined 21 SAS, where evasive driving was part of our training. In the movies, it looks dramatic: loads of screeching and burning rubber. In fact it's a very technical skill, and the best drivers often appear the least spectacular.

Off-road driving is one of those skills that you can keep on improving over the course of a lifetime. However, if you can get to grips with a few basic skills, you'll be well equipped for those adventures that require you to get behind the wheel.

DO A RECCE

It's a lot easier getting *into* a difficult situation than getting *out* of it. Once you're five feet into that mud, or sliding out of control down the loose side of a hill, it can be near impossible to reverse your way out.

In a survival situation, you need to question and study everything in your surroundings. So when you're driving in difficult terrain, it's often a smart move to get out of your vehicle and walk the ground ahead (undramatic as it sounds!). If you feel things getting difficult, sometimes discretion is the better part of valour. Don't plough on; rewind, reverse – retreating exactly back the way you came – then get out and do your recce properly.

With this in mind, it's always best to travel in pairs, and this is especially true with off-road driving. If you have two vehicles (and a good, sturdy tow rope) you can pull each other out if you get into trouble. If you only have one vehicle but two passengers, one of you should drive while the other acts as a 'spotter', guiding you across any tricky obstacles.

TYRE PRESSURE

Off-road driving is all about wheel traction – making sure that the rubber is firmly gripping the terrain beneath it, whatever that terrain may be. As soon as those tyres start spinning, or slipping on unfamiliar surfaces, you've lost control of the vehicle. You can increase your wheel traction in most circumstances by lowering your tyre pressure. This not only gives you better traction but a more comfortable ride too.

The looser the surface, the lower the required tyre pressure, so snow, mud and especially sand would benefit from less pressure. Start by reducing the tyre pressure by one half of your usual road pressure. If the vehicle is still struggling, you can continue to reduce it by 2 psi at a time – small reductions are your friend here.

Remember, though, that the lower your tyre pressure, the slower you need to drive. Low pressure plus high speed equals increased heat and tyre failure. There is also a risk of the tyre separating from the rims. Not good at low speeds, massively dangerous at high ones.

If you're going off-roading, consider taking a portable air compressor with you. These plug into the cigarette lighter of your vehicle and give you the ability to increase and decrease your tyre pressure.

If, despite your lower tyre pressure, you find yourself stuck in mud, or on ice, and your vehicle has rubber floor mats, you can place these in front of the driving wheels to give you that extra bit of traction you need. (A sack or an old rug will have the same effect.)

HOLDING THE WHEEL

One of the most common driving injuries in a crash, even a small one, is dislocated thumbs, because the driver is gripping the wheel with fingers over the top and thumbs underneath when the jolt happens. Better to have your hands at the ten and two o'clock position with the thumbs over the top. Make small movements and don't slide your hands around the wheel.

CROSSING DITCHES

If you find the way ahead barred by a ditch, don't approach it straight on. Not only do you risk the front of your vehicle making contact with the ground and becoming stuck, you will also tend to reduce traction in your two front wheels. Instead, approach the ditch at an angle so that only one wheel is in the depression at a time. This gives you traction on three wheels rather than two.

DESCENDING STEEP HILLS

Descending steep off-road hills can be very hazardous. Your natural reaction might be to hit the brakes in order to maintain a slow descent, but if you lock your wheels with the brakes, you can lose traction and start sliding down the hill. Instead, keep yourself in first gear and allow the engine braking to absorb some of the acceleration. You might need to use your brake pedal a little bit, but only very gently. If there are ruts in the track caused by other drivers, try to follow them – they'll give you the best traction.

DRIVING THROUGH DEEP WATER

Most 4 × 4 vehicles have a maximum 'wading depth'. If you're in a regular car, you don't want to wade deeper than halfway up the wheel. Either way, you're going to need to know the depth of the water you're crossing, and the only way to do that is to measure it. You're going to get wet unless you have a pair of waders in the trunk. Find yourself a good, sturdy pole, get into the water and measure the deepest point. Better to do this and get wet than to plough on and get stuck.

If you're crossing flowing water, be mindful of the strength of the current – vehicles float, and can easily be carried downstream.

If you have a tow rope, it's a good idea to attach this *before* you drive into the water. That way, if you get into trouble, you increase your chance of recovery.

Most bodies of water will have a trough shape: a downward incline where you enter, followed by flat ground, followed by an upward incline where you exit. You should enter slowly and in a low gear to stop the water splashing. Once you're on the flat, you should accelerate slightly. This will form a wave ahead of the vehicle, which has the effect of lowering the water level around the vehicle's engine. It's important not to stall, because as soon as you do, the wave will

settle and it can be hard to regain traction. Once you're ready to exit, reduce your speed: the wave will recede as it hits the bank and cause a big splash if you hit it too hard.

Sometimes, no matter how much know-how you have, you've just got to improvise. I once saw an overland bus get stuck at the bottom of a V-ravine. It needed jump-starting, but this was impossible because of the lie of the land. So they jacked the bus up, tied the tow rope around the wheel, got everyone in the bus to take the far end of the rope and jump-started the engine while the vehicle was jacked up and immobile. That's true survival for you – it's all about ingenuity and resourcefulness.

K.I.S.S.

(KEEP IT SIMPLE, STUPID!)

Boy racers drive inefficiently and noisily.
Special Forces operators do the opposite.

Always recce the ground ahead so you know
what you're dealing with.

It's all about wheel traction.

Reduce your tyre pressure on loose surfaces.

HOW TO
USE AN
ABANDONED
VEHICLE

You'd be surprised how often you come across an abandoned vehicle in survival scenarios. The Australian outback, for example, is dotted with farm buildings and vehicles that are barely used for months on end – think tractors, quad bikes and all-terrain vehicles. The same goes for large swathes of Africa and Asia. And while it goes without saying that you should never break in to, hot-wire or otherwise damage a vehicle unless the alternative is genuinely dire, if you're hundreds of miles from civilization and your environment throws up an abandoned vehicle, it might just be the lifeline you need.

I for one wouldn't hesitate to appropriate a vehicle in the wild if one of my companions was seriously injured, or if I knew that my own life depended on it.

KEYS

Sometimes, simple is good. If you come across an abandoned vehicle, you're going to save yourself a lot of time and energy if you can lay your hands on the keys. In such situations, this is more likely than you might imagine. In remote locations, it often makes sense for people to leave their vehicle keys near the vehicle.

So, before you start breaking into vehicles or hot-wiring, look for the keys. If it's a farm vehicle and there are outbuildings nearby, go and search them thoroughly – and while you're at it, if you see any extra fuel, grab that too: there might be a long drive ahead of you. If not, or you have no luck, look on the vehicle itself. The most common places to hide keys on the actual vehicle are:

- on top of the tyres
- inside the bumper
- inside the fuel pump cover
- behind a sun visor or under seat mats (if the vehicle's unlocked)

BREAKING A WINDOW

To break into a locked vehicle, you'll need to shatter one of the side windows. If you have duct tape in your pack, stick some to the window in an X shape – this will stop the glass shattering dangerously all over the seat and footwell when you hit it.

Use a solid object to break the window. Most people's natural tendency is to hit the glass in the centre. In fact that's the worst place: go for the corners where the glass is weaker. It's also safer. If you go for the centre, your hand and arm can easily follow through into the glass. If you strike the corner, you'll hit the frame of the window and stop. This is a common police technique.

GETTING A VEHICLE STARTED WITHOUT KEYS

If it's a very modern vehicle, you're going to have trouble starting it because of electronic key fobs and up-to-date security systems.

However, the type of vehicle you're likely to come across in many parts of the world and in this kind of situation will be far from modern – we're talking old Land Rovers, tractors, quad bikes. They are a much easier proposition.

THE SCREWDRIVER METHOD

If there are outbuildings nearby and you can lay your hands on a flat-head screwdriver and hammer, this is potentially the easiest way to get an old vehicle started. Imagine the screwdriver is a key. Bash it hard into the ignition with a hammer. If you're in luck, turning the screwdriver will start the vehicle. (If it's hard to turn, a pair of pliers will help you out.)

HOT-WIRING

You're going to need some wire strippers and, if possible, some insulated gloves.

Rip off (or unscrew if you have a screwdriver) the plastic panel beneath the steering column. You'll find a bundle of wires underneath leading to the ignition unit. The ones you want are likely to be coloured red (there should be two of them) and brown (there should be one or two).

Cut the two red wires and strip the ends. Twist the ends of these wires together. (Use the gloves if you have them, or a tool with an insulated handle – don't touch the wires.)

Now cut the brown wire (or wires) and strip the end. These carry a high current, so don't touch them with unprotected hands either.

If there are two brown wires, touch them together. The car should start. Don't let the brown wires touch again.

If there is one brown wire, touch it to the red wires to start the car. Don't let them touch again.

Wrap insulating tape round the exposed wire ends if possible to stop you touching them as you drive.

BUMP-STARTING A VEHICLE

You may have found the vehicle keys, but if the vehicle is in a remote location, there is a good chance that the battery will have drained. If this has happened, you won't be able to start the engine. Bump-starting is a good workaround, but be aware that it only works on a vehicle with manual transmission.

If you're on the flat, you need two people to bump-start a vehicle. If you can get it to a slope, you can do it single-handed. Here's the process:

1. Turn on the ignition.
2. Put the vehicle in second gear and keep the clutch down.
3. Release the handbrake.
4. If you're on a hill, let the car roll. Otherwise you'll need a push.
5. At about 10mph, release the clutch. The engine should start with a 'bump'. If it doesn't, try again.
6. Keep the engine running for at least twenty minutes to recharge the battery.

OTHER WAYS TO USE
AN ABANDONED VEHICLE

There are no keys. You can't hot-wire it or bump-start it. As a mode of transport, it's useless to you. However, it is still a potentially valuable survival resource.

SHELTER
You shouldn't shelter inside a vehicle in a desert environment, but otherwise it's a good way of protecting yourself from the elements.

FUEL
Drain all the fuel from the tank for use in fire-making, by siphoning it out. To do this, you'll need a tube – you'll find something suitable attached to the engine. Get one end into the fuel reservoir and make sure the other end is lower to the ground. Suck the free end

to get the fuel flowing (don't swallow!), and collect it in whatever bottles or containers you can lay your hands on.

WATER
Collect water from the radiator and/or windscreen wiper reservoir. Purify it before drinking (see page 29).

TYRES
Tyres burn well. If you can get the whole thing off, you'll have the basis for a long, hot-burning fire. If you can cut strips of rubber, these act as a good accelerant.

INNER TUBES
Once you've cut up the tyres, the inner tubes can be used as flotation devices (see page 181) or even as bladders to transport water. For this you'll need to cut out the valve, then tie below the hole once you've filled the tube with water.

SEAT STUFFING
If you can cut into the seats, the stuffing can be used as tinder, or to stuff inside your jacket or boots for insulation against the cold.

CABLES
Pull them out and use them as cordage for tying structures and shelters, or in emergency situations for making a tourniquet (see page 414).

GLASS
Smash the lights or windows and use the shards of glass as cutting tools or for starting fire by magnification (see page 38).

MIRRORS
Smash these for cutting tools, or use them for signalling by reflecting the sun (see page 88).

K.I.S.S.
(KEEP IT SIMPLE, STUPID!)

In remote areas, keys are often kept near,
on or even in the vehicle.

Break side windows at the corners, not the centre.

Know how to bump- and jump-start.

If you can't start a vehicle, think laterally
to use it as a survival resource.

HOW TO

SURVIVE

TYRE

BLOW-OUTS

AND BRAKE

FAILURE

Tyre blow-outs – when your tyre punctures at a high speed – and brake failures can be terrifying. But there are ways of handling them that will keep you alive.

TYRE BLOW-OUTS

Tyre blow-outs can be caused by obstacles in the road. But they can also be caused by under-inflated or distressed tyres. You massively reduce the likelihood of getting one if you keep your tyres properly inflated, and avoid kerb-mounting in general driving.

You'll know you've got a blow-out if:

- Your steering wheel suddenly starts to shudder.
- The steering becomes heavy (this normally indicates a slow puncture).
- The vehicle feels like it's being pulled to the left or right (this normally indicates a medium puncture).
- The vehicle swerves violently (this normally indicates a major tyre burst).

If this happens, your instinct will be to slam on the brakes. *Don't do this.* If you have a front puncture, it moves the weight of the car to the damaged wheel. This will make you swerve badly, or will dig the rim into the tarmac and flip you over. If you have a rear puncture it will increase the drag and cause the vehicle to fishtail, possibly sending you into a 360° spin.

Keep a cool head, your foot away from the brake, and:

- Grip the steering wheel firmly with both hands and correct any swerve. Do *not* yank it hard in either direction as this can cause a flip. Just try to keep the vehicle straight.
- Keep your foot on the accelerator. If you feel yourself losing control of the vehicle, accelerate slightly until you regain control.
- Put your hazard warning lights on. Now ease off the accelerator to bring the speed down.
- Only change down a gear (in a manual transmission) if you can do so in a controlled manner. No jerky gear changes and clutch movements that cause sudden drops in speed.
- If the steering wheel is very difficult to control, don't worry about changing gear. Keep a firm grip on the wheel until

you come to a natural halt.

- Indicate to the left when your speed has reduced enough for you to have more control of the vehicle. You're aiming to come to a halt on the hard shoulder or the side of the road.

See page 151 for instructions on replacing a flat tyre.

BRAKE FAILURE

Sudden brake failure is rare, but potentially catastrophic. To bring your vehicle to a halt, you'll need to use a mixture of engine braking and handbraking.

First, lose as much speed as you safely can by simply taking your foot off the accelerator. Dropping gears and using a handbrake at high speeds can cause you to lose control of the vehicle.

To force the vehicle to slow down with a manual transmission, press the clutch, change down into second gear, and bring the clutch up gently so that it acts like a brake. In an automatic, shift to Low, or change down one gear at a time if there's a manual option.

While you're doing this, gently engage the handbrake – don't tug it violently or you'll lock the wheels and cause a skid.

Constantly be on the alert for safe escape lanes where you can leave the road. Try to aim for uphill slopes to slow you down naturally.

K.I.S.S.
(KEEP IT SIMPLE, STUPID!)

If you have a blow-out, avoid your natural
instinct to hit the brake.

Grip the steering wheel and try to resist the swerve.

Slow down gradually before getting to the side of the road.

Brake failure: use your engine and handbraking
and head for an uphill slope.

HOW TO

BUILD

A

RAFT

If you're making a river crossing, see pages 271–6. For a more long-term structure, or for travelling downstream, you might be better off making a raft. Over long distances, they can help you conserve energy, compared to walking along the river bank.

When you picture a raft, you probably think of a bunch of logs lashed together. That's all well and good, but log rafts are quite difficult to make as they are heavy and labour-intensive. It's good to have a few other methods to hand.

As always, remember that in a survival situation these things don't have to look pretty. They only have to be functional. Your raft won't look as neat as the pictures in this or any other book. That doesn't matter. As long as it floats, it can save your life.

Remember also to construct your raft next to the shore. It might sound obvious but they can get heavy when built. I have seen people not do this, and then not have the manpower to move the raft down to the water! Bad waste of time, morale and energy.

TRADITIONAL RAFTS

Here are two straightforward methods that are practical and energy-efficient to construct. For both these methods you need to test the logs first – some will sink, even when they're dead and dried out.

The first uses pressure bars, lashed together as tightly as you can, to keep the logs in place. If you carve a notch into the end of the pressure bars, you'll stop the lashings from slipping off.

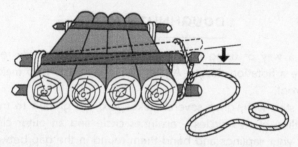

The second uses only two logs, tied so that they have a gap between them of about half a metre. You'll get wet using this construction, but it will keep you afloat and works well if you aren't in a cold climate – sit with your back against one log and your knees over the other.

BRUSH RAFT

A brush raft is principally a construction for floating your pack or equipment downstream, but you can use it as a flotation aid for yourself. Just don't try to sit on it, because it'll probably sink. Think of those floats you use when learning to swim, and use it in the same way.

The idea is to fill a tarp or other waterproof sheet full of light brush and twigs. Wrap the tarp up and tie it securely. The brush inside will create air pockets, so the bundle will float on water.

If you're using a brush raft to float your equipment, you can either put your pack inside the raft, along with the brush, before tying it up, or you can secure the pack to the raft itself. In this case it's worth making a double-layer raft (you'll need two tarps for this) which will give your gear a better chance of staying dry.

DOUGHNUT RAFT

This is a way of using flexible young saplings or sturdy reeds to make a flotation device. Again, you'll need a tarp to make it waterproof.

To start with, drive several stakes into the ground to make the outline of two circles – an inner circle and an outer circle. Collect your saplings and bend them round in the gap between the two circles. They'll eventually create a doughnut shape. Tie the doughnut securely at frequent intervals. When it's secure, lift it from the guide stakes and wrap it in the tarp or other waterproof blanket, which you will also need to tie securely.

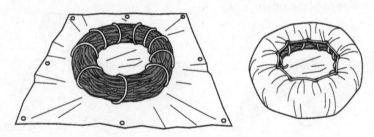

STEERING A RAFT

You can make sails and rudders for your raft, but in a survival situation that's likely to be too time-consuming. Your most expedient means of steering a raft is going to be using branches, either as a punting pole or as an improvised paddle. A branch with a fork in the end works well. Simply tie some clothing or material around the fork to make it into a paddle.

If you can, make a couple and lash one to your raft. They're very easy to lose on the water.

OTHER FLOTATION METHODS

You might not have the time or the necessary materials to make any of the rafts above. If that's the case, you're going to have to use whatever comes to hand.

TROUSERS
You can make a flotation device from your trousers. Take them off and get them wet – they'll hold the air better this way. Tie a knot at the bottom of each leg. Hold the trousers by the waist and swipe them through the air so that they inflate. Scrunch the waistband together before the air can escape, and submerge the trousers, which will keep the air in for longer. A trouser flotation device won't last for long – you'll probably have to keep re-inflating them – but they'll get you across a short stretch of water.

EMPTY CONTAINERS
Anything that holds air will float. Petrol cans, barrels, plastic bottles, waterproof boxes. If the items are small you can lash them together to make more substantial flotation aids. They can also be lashed to the rafts above to increase their buoyancy.

LOGS
You may not have enough logs to build a raft, but a single large log or even a piece of driftwood can keep you afloat.

K.I.S.S.

(KEEP IT SIMPLE, STUPID!)

Rafts don't have to look good. They just have to work.

Brush rafts and doughnut rafts can be quicker and easier
to make than more traditional log rafts.

Make your raft near the shore.

Pretty much anything that floats can be a flotation device.

HOW TO

SURVIVE

IN A

LIFE

RAFT

The sea has taken more lives than we could ever possibly count. It can be a terrifying, deadly environment.

But it is not impossible to survive. Some of the greatest stories of survival have occurred at sea. Men and women have endured, alone and in groups, for months.

In order to survive at sea, you need some kind of flotation device. People die very quickly in the water. Out of the water, you've got a chance. Life rafts are designed to aid survival in emergency situations, but a life raft by itself won't be enough. Your ability to survive will depend on your knowledge, skills and clear thinking – and of course on your will to live.

If you don't have a proper life raft, your chances are a good deal worse, but still not hopeless. Much of this information will still apply.

So this is what you need to know if, having abandoned ship, you find yourself adrift at sea.

LIFE RAFT SURVIVAL KITS

Most life rafts have a survival pack. This will probably include:

- sponges – to bail out sea water, and to absorb rainwater for drinking
- paddles
- a first aid kit
- waterproof torches, plus spare batteries and bulbs
- a signalling mirror
- sea-sickness tablets
- flares
- a repair kit and pump, for re-inflating the raft
- emergency drinking water
- emergency rations
- solar still and/or desalination apparatus
- a sea anchor

These items will be your lifeline over the first twenty-four hours. But they won't last for ever. Your most important survival tool will be the ability to think smart.

SEA ANCHORS

If a plane has crashed or a vessel has capsized, there will likely be a search party. The search will be concentrated on the area of the crash or capsize, but you can drift hundreds of miles a day in a life raft. A sea anchor can slow down your drift.

A sea anchor is not like a regular anchor that goes down to the sea bed. It's essentially a plastic bag with a hole in it, attached to the raft by a rope. It fills with water, becomes heavy, and slows down the raft, a bit like an underwater parachute.

One of your first survival priorities should be to set up the sea anchor to keep you in the search area.

A sea anchor will also stabilize your raft in rough weather, keeping the bow into the wind and waves, and therefore reducing the chances of being broached or capsizing. A sea anchor can

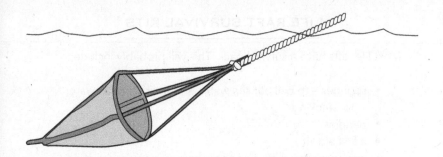

be used to steer, too, so that you drift in the right direction. If it's positioned to the right of the boat, it will steer you to the right of the current; to the left of the boat, to the left of the current.

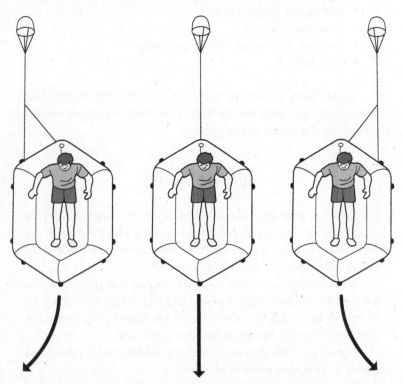

In rough weather, use your sea anchor (or a paddle as a rudder) to head *into* the wind. This reduces the risk of a capsize.

PROTECTION FROM THE COLD

Survival in cold waters is far less likely than survival in warm waters. The cold water itself is a killer – you lose body heat more than twenty times faster in water than you do in air of the same temperature. But the wind is also a big problem, especially if you're already wet. As soon as you get hypothermic (see page 442), it's incredibly difficult to warm up again. You'll stop thinking straight and your ability to survive will be severely diminished. You need to do everything you can to mitigate the life-sapping effects of the cold, so:

- Try to dry all wet clothing. If this is impossible, squeeze out as much moisture as you can.
- Put on as much extra clothing as possible, but try to keep your garments loose and comfortable.
- Keep the floor of the raft as dry as you can.
- If possible, cover the floor with canvas or material to insulate it against the cold.
- Most life rafts have a covered shelter to keep the sea and the elements out. If yours doesn't, try to rig something up to keep yourself protected from wind and waves.
- If there's more than one of you, huddle up. Shared body warmth saves lives.
- Move your body gently to keep your blood circulating and to ward off frostbite.

PROTECTION FROM THE SUN

Warm waters come with their own set of problems. If you're exposed to the sun for too long, you're dead. So:

- Use the life raft's shelter. Again, if it doesn't have one, rig one up with whatever you have to hand. It's essential that you protect yourself from the sun's rays in any way possible.
- Remove excess clothing, but keep the body covered.

- Especially, keep the head and neck covered.
- Dampen your clothes during the day, but make sure they've dried out by night, when the temperature can drop dramatically. However, don't overdo this. If your clothes become too saturated with salt, it can cause dangerous weeping sores all over your body.

WATER

As always in an extreme survival situation, drinking water is going to be one of your biggest priorities. You can't drink sea water as it will simply dehydrate you much faster than not drinking anything and it will cause rapid death due to salt poisoning and kidney failure. Some survivors, however, report having moistened their lips with sea water, and say that it has helped their morale. And in a long-term survival situation, morale is important.

Your approach to water should be two-fold. You need to *preserve* it and you need to *collect* it. When it comes to preserving water, remember that:

- You lose a massive amount of water when you sweat. Even more reason to keep cool in hot climates.
- Vomiting drains you of water. If you have sea-sickness tablets in your survival pack, use them.
- You lose a lot of water digesting food. If water is scarce, don't eat.

There are several ways of collecting water on a life raft.

1. A life raft survival pack will probably contain a solar still. They work on the same principle as the one shown on page 28, except you put salt water in the base, which condenses and collects in the cup. You should set these up as soon as you find yourself in a survival situation, but remember that they can only be used if conditions are calm, otherwise salt water will mix with the distilled water.
2. When it rains, get to work, and fast. Many life rafts have

pouches in their canopy for catching rainwater. If not, collect it in any and every container you have. Spread tarpaulins or canvases in a bowl shape over as big an area as possible (if they are encrusted in salt, wash them in sea water first).

3. Collect dew by leaving a tarp or clothing out at night. If you erect it like a tent and fold the edges up, dew will collect in the creases. Even small amounts can be life-saving.

4. If you manage to catch a fish, you can get fresh water from it – see page 104.

FOOD

Your main source of food will be fish. There are poisonous fish, but the rule of thumb is that in the open sea most fish are edible. The most dangerous ones, like the deadly stone fish, are often found close to shore in shallow water. (All fresh-water fish, however, are edible.) Remember, though, what I said above: if you have insufficient water, don't eat. The process of digestion will dehydrate you further.

To catch fish while adrift at sea you can do any or all of the following:

- If there's fishing equipment in your survival pack, use that.
- You can improvise fishing lines using your shoelaces, or by unravelling threads from your clothes or any other material. Tie them together in short lengths of four or five threads.
- You can improvise hooks from hair pins or from pieces of driftwood.
- If you have a knife, you can tie it to the blade of an oar, or a piece of driftwood, to make a fishing spear for larger fish.
- If you have a light, use it at night to attract fish.
- During the day, fish are attracted to shadow, so fish in the shadow of your raft.
- If you catch a fish, use its guts as bait.

- If you catch fish in a hot environment, cut strips from its flesh and hang it up to dry – it will keep good for several days (see page 117).

Another source of food will be sea birds. Your only real chance of catching them is if they land on your raft, and this does happen. One of the greatest stories of survival at sea is that of Louis Zamperini, who, adrift for forty-seven days, managed to catch albatrosses and other sea birds with his bare hands and survived by drinking their blood, devouring their flesh and drinking rainwater he caught in a tin can. Another famous sea-survival trip is Captain Bligh's 4,164-mile journey in an open 23-foot launch after the mutiny on the *Bounty*, during which he and his men ate boobies, so called because of their habit of landing on boats without a care in the world – deemed by mariners of the time as a rather stupid thing to do.

A good way to catch sea birds is by using fishing line made into a noose, baited with fish guts. Lay the line on the roof of your raft and always be ready to pull the noose tight if a bird lands. Or you can bait a hook, tied to a line and put it on the outside of your raft. A bird will take the bait and get the hook caught in its throat as it tries to fly away.

If you manage to catch a bird, use its guts and feet for bait and its feathers for insulation, or as lures for catching fish.

Your final potential source of food is seaweed. It's mostly edible, but it's very salty. Only eat it if you have plenty of fresh water.

ROUTINES

If there is more than one of you on a life raft, routines are important. They keep you motivated, and your mind occupied. Don't underestimate the importance of morale in a survival situation. Jobs might include:

- bailing out water
- first aid
- food prep

- water rationing
- paddling
- navigation
- look-out duty

This final job is crucial. There should be somebody scanning the sky and the horizon *at all times* for an aircraft or a ship. You might only get the briefest glimpse. When that happens, you need to signal for help.

SIGNALLING

You are likely to have two types of flare in your survival pack: hand flares and rocket flares.

Hand flares are visible from a distance of about three miles, so they should only be used if you can clearly see another boat, aircraft or people on the shore. If you can't, don't waste them, no matter how tempting it might seem or how much you might be panicking. Once they're gone, they're gone. Red flares can be seen best at night, orange flares can be seen best during the day. If you see potential help, don't let all your flares off at once. Wait a few minutes to see if a vessel changes course to head in your direction, before igniting a second one.

Rocket, or 'parachute', flares can be seen from about 25 miles away. Use these when a vessel is too far away for a hand flare, or when you think you might be within 25 miles of land. Again, only use a second flare if there's no evidence that your first flare has been seen. If a vessel steers towards you because it has seen your rocket flare, now use a hand flare to pinpoint your location.

Survival packs also sometimes include signalling mirrors. See page 88 for instructions on using these.

SIGNS OF LAND

There are ways of telling that land is close – so you can steer towards it, or use a parachute flare – even when you can't see it.

- Look out for flocks of birds, morning and night. In the morning they are probably heading away from land. In the evening they are probably heading towards it.
- Cumulus clouds often form over an island.
- The sea will carry increased amounts of vegetation and/or driftwood.
- The water will be of a lighter colour.

K.I.S.S.
(KEEP IT SIMPLE, STUPID!)

At sea, dehydration, extreme cold and extreme heat are what will kill you. Be prepared.

Preserve and collect water at every opportunity.

Don't eat if you have no water.

Only use your flares when you know there's a chance of them being seen.

HOW TO

CLIMB

IN A

SURVIVAL

SITUATION

Climbing a rock face unaccompanied and unroped is difficult, dangerous, and you only get it wrong once. If you can go *round* a rock face, it's always better to do that. But if your life depends on getting *over* it, there are some key ways of maximizing your chances.

PLAN YOUR ROUTE

In almost every survival situation, time spent planning is seldom wasted. But it's particularly true of survival climbing. If you watch professional climbers, you'll see that they spend ages at the bottom of a rock face working out their route. They also know that climbing down is *far* harder than climbing up. When you're climbing up, you can *see* where you need to put your hands and feet. Climbing down, you have to feel for purchase. I've done some gnarly climbs that haven't gone the way I've wanted, and I've chosen to carry on by traversing via an alternative route rather than go back down. You need to plan your ascent before you commit yourself.

KNOW WHAT TO LOOK FOR
As you plan your route, you're looking for:

- cracks – vertical as well as horizontal
- ledges
- big features
- less steep areas
- solid rock

Picture yourself on the climb. Where are you going to put your feet? Where are you going to put your hands? Of course, you'll need to be adaptable, but don't fool yourself. Mentally put yourself on that cliff and plan every stage of the climb.

KNOW WHAT TO AVOID
This is as important as knowing what features to look for. So, avoid:

- wet areas – water and rock do not mix
- green areas, which are likely to be slimy – potential death traps
- loose areas of rock or gravel
- overhangs

CLIMBING ISN'T ABOUT ARMS

At least, not principally. Four other factors are going to keep you pinned to that wall:

1. **Balance.** Keep your body close to the wall. Too many people arch their bodies, with their bums sticking out, when they try to climb. That's going to exhaust you very quickly as the weight goes on to your arms rather than your legs. You can occasionally push your body out to look up, but generally you should be pressed right up to the wall.

2. **Core stability.** You need to keep your core muscles tense. Really focus on this, as it will keep your limbs and hips in the proper alignment.

3. **Legs.** Your legs have massive muscles, significantly stronger than your arms. They are much more efficient tools for getting up the rock face.

4. **Nerve.** Try to stay relaxed and calm. If fear takes over, you'll start to shake and burn through the adrenalin, which will fatigue you fast. I've had it a few times when climbing, right at that difficult move. In the climbing world it's called 'disco leg'.

Of course, now and then you're going to have to 'finger jam' into tight crevices or cling to small nodules in the rock with your hand and use your biceps to pull yourself up. Be aware, though, that your hands are going to hurt like hell after being squeezed into cracks and gripping rough rock. Use them mainly to balance, and your legs to climb.

CONSERVE ENERGY AND ACCEPT RISK

If you're using your legs and core to get up the rock face, you'll be using a lot less energy than if you use your arms. This is important. Climbing is exhausting.

You don't need to go fast up a rock face. Accept that this is a slow process, and climb within your own level of physicality. Take your time. When people get scared they tend to rush and then burn out. Pacing yourself is key.

However, you also need to understand that the longer you're *on* that rock face, the more exhausted you're going to get. So, find a rhythm. Keep moving as much as you can. Most importantly, know that the longer you hold on to the rock face, looking for your next move, the faster your energy is going to drain. It's human nature to stop as you try to muster the courage to go for a big move. But if your situation isn't going to improve by waiting, you're much better advised to make a quick decision and turn it into an action.

In fact, the ability to quickly turn difficult decisions into well-measured actions is one of the core skills of an adventurer. It is about knowing how to balance risk. When to push on and when to retreat. It is called survivor's wisdom. It takes time to learn, but essentially it comes from experience and having a calmness about you so that you can listen to that inner voice.

Another human reaction to fear is to freeze. Both rushing and freezing will increase your chances of making a bad move. Instead, aim for a balance between considered moves and constant progress, against not burning out or making hasty, bad decisions. If in doubt, take a moment, think it through, then commit and go for it. There's no power in tentativeness. You need ultimately to go for it. The only thing worse than a bad decision is no decision.

Ultimately, don't be too proud to retreat. Retreat isn't failure. In climbing, retreat is living long enough to have another attempt. The greatest feats of climbing have often taken years to achieve. Patience is important.

CHIMNEYING, STEMMING AND MANTELING

These are three emergency climbing techniques that can get you up a rock face faster and smarter.

Chimneying is a means of climbing up two opposing rock faces, or any vertical crack big enough to climb inside. The idea is to use opposing pressure to stop yourself falling. To do this:

1. Start with your back against one wall.
2. Place one hand and one foot against each wall.
3. Straighten your legs to push upward.
4. Re-establish your hand position.
5. Move your back leg so it's adjacent to your front leg.
6. Move your front leg to the back position.
7. Repeat.

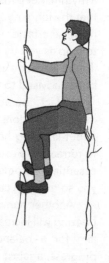

Stemming is an alternative to chimneying, and faster if you get the hang of it. You face forward and spread your arms and legs out in an X shape, then shuffle up the rock face.

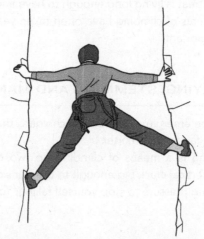

Manteling is the technique you need in order to get on to a ledge or over the top edge of a cliff.

1. Get your chest to the level of the ledge using your arms and legs.
2. Get your heel over the ledge.
3. Push with your hands and elbows to raise your body to ledge height.

At the end of the day, remember: this is not a climbing competition. Your climbing doesn't have to look tidy or cool. It is true that in climbing tidy is often also efficient, so you should aim to climb in a way that is smooth and combines your big muscle groups and your balance in order to overcome. Ultimately, though, this is about survival. Using your knees, for example, might be frowned upon in climbing clubs, but in our scenario you have to use any technique you can to get yourself up and over the rock face, alive and to safety.

K.I.S.S.
(KEEP IT SIMPLE, STUPID!)

Plan your route up a rock face in detail before starting, and include a descent back-up plan – just in case.

Avoid wet, green and loose areas.

Balance. Core stability. Legs. Nerve.

Pace yourself and move efficiently.

Know when to push on and when to retreat.

HOW TO

ABSEIL

IN A

SURVIVAL

SITUATION

When most people think of abseiling – or rappelling (they're the same thing) – they imagine someone in an elaborate harness making their way down a vertical cliff face or building.

Well, that *is* abseiling, but that's not *all* there is to it. Abseiling is simply a way of using a rope to move down difficult ground. That might be vertical. More likely, though, it'll be a steep incline that is too dangerous to travel down unaided.

Abseiling is a big thing in the special forces. Soldiers learn to do it so they can rappel out of helicopters or down buildings. To do this, they use a special harness with a variety of descender devices, such as a figure of eight, which is simply two solid metal circles, to attach themselves to the rope. They use sturdy rappeller's gloves to protect them from the damaging effects of friction burn, and they have all the high-end gear at their disposal to slow their descent or to stop themselves continuing to fall if they get injured.

That's all great, but in a survival situation you'll have no special equipment. No harnesses, no figures of eight, no gloves. Knowing how to perform a perfect figure-of-eight rappel will be completely useless.

Instead, I'm going to show you three abseiling techniques, developed by pioneering mountaineers long before the invention of harnesses, that use nothing but rope and body friction. I'm not saying that a figure-of-eight abseil isn't safer, easier and more comfortable. It is. But these are the abseils that could save your life. They can be performed with almost any bit of rope you manage to get your hands on, as long as it's strong enough. You can even do these abseils using a sturdy length of jungle vine, if that's all you have.

TYING OFF

But before you can rappel, you need to attach your rope to a fixed point at the top of your slope. You can do this in two ways.

First off, you can tie your rope to a tree. I would use a secured bowline for this (see pages 67–8). Or remember our knot mantra: if you can't tie knots, tie lots. Any knot will do, so long as it doesn't slip, and the more knots you tie, the safer the rope is going to be. Remember, this isn't technical climbing. This is just for getting the job done and staying alive. Tie ten overhand knots if you have to. Just do whatever you can to make sure that rope is going nowhere.

But there's a problem with tying a knot firmly to a fixed point at the top of a slope: once you've rappelled to the bottom, you'll have to leave it there. If you're in a survival situation, this is a very bad idea. You never know when that rope is going to come in useful. You can't re-use it if it's tied to a tree a hundred feet up a vertical rock face.

By far the better option, therefore, and assuming your rope is sufficiently long, is to loop it round a fixed point, be it a tree or a boulder, so that you effectively have a double length of rope in your hands. In the diagrams below, you'll see that's what we have.

It's worth remembering that if you have to rappel down a snow-covered slope, you can improvise a fixed point at the top of the slope by creating a snow bollard. These only work where the snow is hard-packed – powder is no good. Using a knife or ice axe, dig a horseshoe shape in the snow or ice around which you can wrap your rope. The sides need to incline inward towards the ground so the rope doesn't slip off the bollard. Make sure the snow is packed down as much as you can before threading the rope around the bollard, and you can even pack snow on top of the rope to secure it further. Double-check the bollard first, by putting all your weight on the rope and pulling, before you commit your life to it.

Snow bollard

Rope

Horseshoe trench

ANGEL WINGS

This technique is good for short, low-risk inclines of up to about 45°.

Put your arms out in a crucifix position. Hold the top end of the rope in your non-dominant hand. Pass the rope across your back – or even better your rucksack as this provides extra friction and stops you getting rope burn. Twist it round the wrist of your dominant hand, which should be pointing down the slope, as shown, to give extra friction. You should use this hand to control the speed of your descent.

CLASSICAL ABSEIL

This is also known as the Dülfersitz after the guy who invented it, a German mountain climber called Hans Dülfer. It's good for slightly steeper slopes, and it's a basic safety skill for most mountaineers.

The classical abseil is uncomfortable. You're going to hurt by the time you get to the bottom, but that's a small price to pay for being alive.

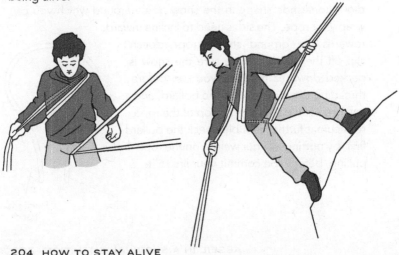

Facing uphill, straddle the two lengths of rope coming down from the anchor. Pull them through your legs and around your hip. Pass the ropes across your front, over your non-dominant shoulder, behind your neck and along your dominant arm.

You need to descend backwards, your knees bent and your feet at shoulder width apart. Use the rope in your dominant hand to alter your rate of descent, but never let go of it as that will cause you to fall.

SOUTH AFRICAN ABSEIL

This can only be done if you've looped your rope around an anchor rather than tying it, because you need two strands. This is for even steeper slopes than the classical abseil. It's a bit more comfortable, though, because the rope cuts into your body less, and you only need one hand to control it.

Facing the anchor, run a length of rope under each arm. Cross them over round your back and bring them to the front of your body. Now, pass both ends under your crotch and hold them behind your back. As you descend, always keep one hand on the rope behind you as this will control the speed of your descent.

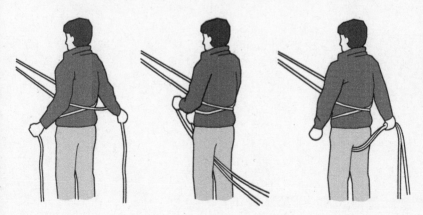

K.I.S.S.
(KEEP IT SIMPLE, STUPID!)

Abseiling is for moving down dangerous ground,
not just vertical cliffs.

Forget about technical abseiling – in a survival
situation you won't have the gear.

Loop your rope rather than tying it – it's a precious
resource and you want it back.

Choose your abseiling technique according
to the steepness of the slope.

HOW TO
LAND A
HELICOPTER
IN AN
EMERGENCY

Flying a helicopter is hard. Non-pilots have a very slim chance of getting to the ground safely. But here's how to increase those chances if you find yourself forced to take the controls of a chopper.

Before you get into any helicopter, there are some things you need to know.

- Whenever you exit a helicopter, move forward. The pilot should always tell you to leave the area between the angles of ten and two on a clock face, if the nose of the helicopter is at twelve. If you don't walk forward, you risk walking into the tail rotor. People have done this even in a non-emergency situation. It's a messy way to die.
- Be aware when exiting a helicopter on uneven ground such as in the mountains. The blades are only ever a few feet above your head and if the ground is sloping upward then you are in danger of losing that head. Staying tight in and close to the helicopter as it lifts off is often the safest place to be. This is especially true when landing at non-designated landing zones.
- Whenever you get a safety briefing on a helicopter, find out how to switch the engines off. It's usually a big red button with a cap over it so that you don't engage it by accident. Likewise, ask where the EPIRB emergency locator beacon is fitted. The reason you need to know these two things will become clear in a minute.

HOW A HELICOPTER FLIES

There are three main controls on a helicopter. They are:

- **The collective.** This is a lever, rather like the handbrake on a car. It's normally positioned to the left-hand side of the pilot's seat. It controls the angle of the main rotors, and so the lift of the helicopter. Pull it up, and the helicopter goes up. Push it down, the helicopter goes down.
- **The pedals,** left and right. These control the angle of the tail rotor blades. The pilot uses them to rotate the chopper left and right: left pedal to make it face left, right pedal to make it face right.
- **The cyclic.** This is normally positioned between the pilot's legs. It looks like a big joystick. It adjusts the main rotors in a different way to the collective. The helicopter will move in the direction in which you move the cyclic – forward, backwards or sideways.

All this sounds simple, but it's not. Controlling the collective, the pedals and the cyclic at once is tough. It takes new pilots a long time to get the hang of it. If this is the first time you've done it, the chopper will most likely be yawing violently. To increase your chances of getting to the ground safely, you should:

- Keep your eyes on the horizon, rather than directly in front of the helicopter. This will help you keep the aircraft level.
- Use very gentle movements on all the controls.
- If possible, turn the helicopter to face into the wind – look for trees or flags on the ground to tell you which direction it's blowing. This will make it easer to land.
- Gradually lower the collective to bring you down to ground level.
- As you ascend or descend the helicopter will start naturally to spin – this is yaw. You need to counter this with the pedals.

AUTOROTATION

This is a way of landing a helicopter in the event of an engine fail. Autorotation uses your downward fall to move air over the main rotors and slow your descent. You'll still only have a matter of seconds before you hit the ground, so you need to think and act fast.

- Immediately move the collective to its lowest position. This will force the chopper to descend.
- Use your right pedal to correct the sudden yaw of the aircraft and get you facing in the direction of flight. (This assumes your rotors are turning anticlockwise. If they turn clockwise use your left pedal.)
- Pull back on the cyclic to level the nose of the chopper.
- When you're about 50 feet from the ground, you need to 'flare'. Pull the cyclic back gently to raise the nose of the chopper and reduce the speed of descent.

CRASH LANDING

If you're doing either of these procedures for the first time, you're probably going to have a crash landing. The nose will likely be forward and the rotors will have stopped because they'll be in contact with the ground. But the engines will still be roaring and overheating. You need to kill them before they ignite any fuel. That's why you need to know how to switch the engines off. It's worth spending a few seconds finding and hitting that red button before you leave the aircraft, otherwise there's a high chance of an explosion.

Choose a clear path, away from the danger of spinning blades, before you exit. Now get out of there quickly. Be aware, as always, of the tail rotor.

LANDING IN WATER

Most helicopters, when they hit the water, will turn upside down. Before impact, be ready to open the door as soon as you touch down, before you go under. Remind yourself how to release your buckles and seat belt, too. Be ready to exit the aircraft.

Always know where your helicopter EPIRB is located. This emergency locator beacon is generally yellow in colour and the size of a small lantern. This should be activated as fast as possible. It sends a distress signal with your location so you can be found.

When landing in water, the skid floats should automatically inflate to keep the helicopter from sinking. Anticipate them failing and be ready to evacuate fast.

Before you leave the helicopter for good, and while it is still afloat, aim to take with you the EPIRB, life raft and emergency grab-bag that all helicopters should have if operating over water. This should include a waterproofed handheld VHF radio, flares and drinking water.

Remember: if you have a life jacket on, don't inflate it until you're out of the aircraft, otherwise it'll just pin you to the top and you'll drown.

Note: if the door fails to open for any reason – because it has buckled on impact, for example – most choppers have straps on the windows that you can pull to pop them open and make an escape route.

Either way, always make a point of noting your possible escape routes before flying in a chopper over water.

K.I.S.S.
(KEEP IT SIMPLE, STUPID!)

Know where the engine kill switch and EPIRB are.

Collective: up and down. Pedals: rotate left and right.
Cyclic: forward, backwards and sideways.

Land into the wind and flare.

Exit a chopper between ten and two o'clock.

Crash landing in water: door open, seat belt release, grab the
safety equipment, swim away asap. (If it's a choice
between living and grabbing gear, then live.)

HOW TO

FLY A

PLANE

IN AN

EMERGENCY

You're on a commercial aircraft. Both pilot and co-pilot are incapacitated. What do you do?

GET IN POSITION

There will most likely be two seats in the cockpit. You want the left one – It gives you easier access to the controls you'll need, although you can fly from either side. If the pilot is slumped over the controls, move him out of the way. Your survival priority is controlling the plane, not tending to the pilot. Strap yourself in.

If the aircraft is cruising, you're almost certainly in autopilot mode. If so, the aircraft will be flying straight and level. Don't touch the flight controls just yet.

FLY LEVEL

If you're not in autopilot, you're going to need to level the aircraft. The steering wheel on an aircraft is called the yoke and it will be right in front of you. Grab the yoke. If you are in a dive, pull back to level off. If you are ascending, push down. If you are in a steep turn, rotate it left or right to straighten up.

If you can't see if you're flying level or not, you need to consult the altitude indicator. It looks like this:

The horizontal lines in the centre represent the wings of the aircraft. The top half of the indicator should be blue, the bottom half brown. Use the yoke to bring the wings in line with the artificial horizon.

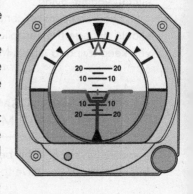

The autopilot button on an aircraft is normally positioned in the middle of the control panel. It will be marked 'Autopilot', 'Autoflight', 'AFS' or 'AP'. Don't press it just yet: it's not as simple as flicking a switch and having the plane fly itself. Your priority is to make a Mayday call so you can be helped to put the plane safely on the ground.

MAYDAY CALL

There are normally two ways to make a radio call: over the pilot's headset or using the handheld radio to the left of the pilot's seat. You should use the handheld radio, because the headset radio requires you to press a button on the yoke that you might mistake for the autopilot disconnect button.

The radio will have a push-to-talk button. Make sure you release it when you want to listen.

Make your call: state 'Mayday Mayday Mayday'.

If you receive no response, you need to change the frequency of the radio to the international air distress frequency. This is 121.5 MHz. Repeat the Mayday call.

This will alert air traffic control to the fact that you're in an emergency situation. Explain to them clearly, in English and without attempting any kind of jargon, what has happened.

Now do exactly as you're told. Most modern airliners have the capability to land themselves, but will still require pilot input. You will probably be talked through certain manual procedures as you are coming in to land, depending on the conditions and which aircraft you are flying.

If you can't make contact with anyone over the radio, you should engage the autopilot to keep the plane flying level while you keep trying to raise a Mayday call.

LANDING A SMALL AIRCRAFT WHEN YOU CAN'T MAKE A MAYDAY CALL

If you've never done this before, your chances of survival are slim, but here's what you have to consider.

GOOD EMERGENCY LANDING SPOTS
If you can't land at a runway, you need to locate a large patch of flat, unobstructed ground. Think beaches, golf courses, flat farmland or deserted roads. The aircraft will hit the ground at speed and continue moving for a long time. You'll need 500 metres minimum of landing space, although some small planes can land on a dime.

You always want to land *into* the wind if possible – a crosswind landing is significantly more difficult (this is why major airports have multiple runways at angles to each other, so the correct runway can be selected according to the prevailing wind). As you get closer to the ground, look for indications of wind direction such as trees or flags. Turn your steering wheel or yoke left or right to turn into the wind.

GETTING TO THE GROUND

Locate your airspeed indicator. It will look like this:

You make the aircraft descend by pushing down on the steering wheel or yoke. However, this also makes the aircraft speed up. So, you need to look for a black knob labelled 'throttle'. Pull on it to reduce the engine's power and maintain a stable speed. (This is a bit like taking your foot off the accelerator of a car when travelling downhill.)

Once you have located your emergency landing spot, you need to slow the aircraft down. Pull the throttle about halfway back while gently pulling back on the control wheel.

Now, focus on the green arc on the airspeed indicator. The lower end indicates the speed at which the aircraft will stop flying. On a light aircraft, it's normally around 70 knots. Add about 30 per cent – so about 90 knots if the bottom of the green arc stops at 70. This is your target speed.

Look for the landing gear switch. It will probably have a picture of a tyre on it. Engage it. If you can't find it, don't worry – you're aiming to survive this landing, not to make it pretty.

Lower the nose slightly, but not so much that the airspeed becomes too high. Line up your target landing site with a fixed point on the windshield. If the target point seems to go below the plane, you're flying too high: reduce the throttle and lower the aircraft's nose. If the target point seems to rise, you're flying too low: increase the throttle and raise the aircraft's nose.

As you're about to hit the ground, pull the throttle right down and pull back on the steering wheel a little to stop the nose from dropping. When you hit the ground you'll probably bounce. Hold the nose steady and wait for the aircraft to come to a halt.

If you see a red knob marked 'mixture', pull it. This will stop the engine. Now get out of there.

K.I.S.S.
(KEEP IT SIMPLE, STUPID!)

If the aircraft is flying level, don't touch anything.

If the autopilot isn't engaged, use the altitude indicator to resume level flight.

International air distress frequency: 121.5 MHz.

State 'Mayday' and follow instructions.

HOW TO

PARACHUTE

IN AN

EMERGENCY

I've done a significant number of freefall jumps over the years. Of those, 99.9 per cent have been fine. But as with any emergency situation, things only have to go wrong once for the results to be disastrous. For me, that one occasion occurred when I was skydiving with friends in southern Africa in 1996. I engaged my chute at 3,000 feet, travelling at terminal velocity (about 120mph). I slowed down, but when I looked up I saw that my chute was tangled. I tried to control it, but before I knew it, I hit the ground like a rag doll. The result was a fractured spine and a hospital nickname of the 'Miracle Kid' for having avoided being paralysed.

The takeaway message? Parachuting is normally safe. When it goes wrong, the results are often fatal. But not always.

Here's what to do when you're forced into an emergency parachute jump.

FALLING STABLE

Before you even consider engaging a parachute, you need to be falling stable. That means your body must be symmetrical. Your arms and legs should be spread out, with your arms above your body and your legs pointing up at a 45° angle. Your head should also be above your body, and you should be relaxed in the air, falling belly first. If you're twisting around in any way, there's a high risk that the parachute will tangle.

The aim is to be falling like a shuttlecock, with your hips being the centre of gravity. If you aren't arched like this, you will flip over on to your back. For the untrained, it can be hard to get spun round again to be falling face first. If you try to deploy a parachute when inverted, there is an even higher risk of becoming tangled in the lines.

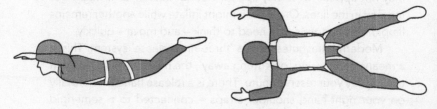

DEPLOYING THE CHUTE

WHEN TO DO IT

If it's an emergency situation and this is your first time, you should do this as soon as possible. If you're freefalling from above 20,000 feet, however, you have an added complication. At this altitude there is scant oxygen – that's why special forces paratroopers wear oxygen masks for high-altitude jumps. Without this extra equipment, you risk suffering oxygen starvation (hypoxia) which can result in lack of consciousness and an inability to deploy your chute.

So if you deploy too high, it will take you longer to get to a lower, oxygen-rich altitude. It's a Hobson's choice – you're in a desperate situation either way – but on balance I would still recommend deploying your chute as soon as possible. You might survive oxygen starvation for a few minutes; you probably won't survive hitting the ground at 120mph.

HOW TO DO IT

Rip cords are pretty much a thing of the past. Most modern parachute rigs have a pilot chute, or 'drogue', to initiate deployment. This is a miniature parachute situated in a small pouch at the bottom of the rig. You simply pull it out and throw it out and to the side, into the main flow of the slipstream. The drogue chute initiates deployment of the main chute.

If you let the pilot chute just 'dribble' out of your hand, there is a chance it will get caught in the 'bauble' of still air on your back, and will never get into the main slipstream that will in turn pull out the main chute.

WHAT IF IT GOES WRONG?

It doesn't happen very often. But it does happen. The main canopy may not appear. Or it may appear but not inflate. Or it might get tangled in the lines. Or part of it might inflate while another remains limp. If this happens, you need to think – and move – quickly.

Modern parachutes have a 'three-ring release' system. This is a means of releasing, or 'cutting away', the main chute so that you can deploy your reserve chute. There is a release handle – normally on your right-hand shoulder straps – connected to a semi-rigid cable that activates the three-ring system and allows the faulty main chute to be jettisoned. You'll need to give it a good tug.

Once you're free of the main chute, two things might happen. On some parachute rigs, your reserve chute deploys automatically when you cut away. On others, you'll need to pull another handle to deploy it manually. The reserve handle is normally on the opposite side of the cut-away handle, on your left shoulder strap. It will probably be a loop made of either red cloth or metal.

SCOPING YOUR LANDING ZONE (LZ)

Let's assume that either your main chute or your reserve chute has deployed and you're falling safely towards earth. You need to scope out a safe LZ. You're looking for a wide open, flat area, free of trees, water or other obstacles.

It might be that you need to steer your parachute in order to position yourself safely. Most modern (square) parachutes allow you to do this by using the two steering toggles, or 'brakes', on either side of the rig. Pull the right-hand one and you'll turn to the right, pull the left-hand one and you'll turn to the left.

You can also pull both toggles so that the chute 'flares'. This will slow you down considerably. Unless you're an expert, you should avoid doing this at altitude because it can make the chute collapse. Applying a little brake to both toggles, though, will slow your forward speed if you are concerned about over-flying your LZ. (Put your hands and brakes back, fully up again, before landing, to gain speed to allow you to flare fully and slow down for the actual touchdown – see below.)

LANDING

You should always aim to land *into* the wind, because it will act as a brake, making the landing slower and safer. This, in turn, will make you less likely to injure yourself. A broken leg or ankle in the wild can be just as dangerous as a failing parachute.

Landing safely, though, is easier said than done.

As you descend, you'll be able to assess which way the wind is blowing if you observe your speed over the ground below as you turn in different directions. Also, flying into the wind can be noisier as the air flow is greater.

To steer into it, pull one of the toggles, and keep it pulled until you've turned a half circle. Don't do this too close to the ground – you definitely don't want to be turning as you hit earth, otherwise you will hit too fast and injure yourself. When you perform any turns, you lose height significantly.

At 100 feet minimum you want to be steering in one direction. Do *not* execute a low turn, even if you realize you are not heading into the wind. Better to take a downwind, slightly faster landing than try to hook-turn too late and smash into the ground.

Unless you're an experienced parachutist, you'll probably be hitting the ground at quite a speed. The parachute landing fall (PLF) can minimize the impact of the fall on your body. Here's how you do it:

1. Press your feet and knees together. Bend your knees slightly. Point the balls of your feet forward. Tuck your chin against your chest.

2. When your feet touch the ground, twist your upper body in the opposite direction to that in which you're drifting. Push your outside knee into your inside knee. Fall in the direction of the drift.

3. Twist your body to expose your calf and thigh muscles. Keep your neck muscles tense, your chin against your chest.

4. Roll in the direction of the drift, keeping your elbows in and your hands up to protect your face.

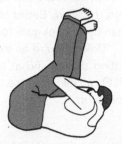

WHAT IF YOUR RESERVE
CHUTE FAILS TO OPEN?

If this happens, your chances of survival are tiny.

But they're not non-existent. There are a number of stories of people falling from planes and surviving the impact. If it happens:

- Adopt the falling stable position described on page 221. This will reduce your falling speed.
- In this position, you can steer your body by lowering your left or right elbow and leaning in the direction you want to turn. You can move backwards by pushing your arms forward and bending your knees. You can move forward by putting your arms to your side and extending your legs.
- If you're falling alongside another person who has not yet deployed their parachute, signal to them that you're in trouble. Try to manoeuvre yourself towards them. Link hands. Now, before they engage their chute, hook your arms through the straps of their rig and wrap your legs around their hips, below the chute. Otherwise you'll never be able to hold on to them when their chute is engaged.
- If you're on your own, you need to aim for terrain that will give you a soft landing. In practice, that means forested areas, marshy areas, or snow-covered areas.
- Avoid buildings, and definitely avoid water: at speed, it's like hitting concrete.
- It helps if you can land on the side of a hill, because the slope helps absorb some of the momentum.
- Make yourself as wide and big as possible to trap as much air in your clothing to slow your rate of fall. Head down or feet down and your fall rate will speed up dramatically.
- Try to use the parachute landing fall even in this situation.

K.I.S.S.

(KEEP IT SIMPLE, STUPID!)

Fall stable and deploy your pilot chute.

If the main chute fails or is uncontrollable, go for the cut-away handle and reserve chute.

Steer to a safe LZ.

Land into the wind and adopt the parachute landing fall.

TERRAIN SURVIVAL

HOW TO

SURVIVE

IN THE

DESERT

The desert might just be the most hostile place on Earth. Exposure to intense heat along with a lack of water can kill you in a matter of hours. Even if you stay alive, you'll be prone to

- extreme constipation
- pain when urinating
- heat cramps and rashes
- sores in your armpits, on your groin and between your toes

All these affect your ability to think clearly and move swiftly. So, your two survival priorities are shelter/shade and water.

SHELTER

If you are travelling through the desert in a survival situation, you should rest by day and move by night. Daytime temperatures can top 50°C, but sometimes drop close to zero at night. This means that moving at night-time isn't necessarily easy, but trying to move during the day if you're not used to it is pretty much impossible.

CLOTHING AS SHELTER

Your clothing is your first line of defence against the sun. Keep your skin covered, but try to keep your clothing loose. Most importantly, you need to keep your head covered. It acts like a solar panel in the sun. If it's unprotected, you limit your ability to survive in a desert dramatically. If you don't have a hat, a scarf, a T-shirt or any piece of material will do.

Urinating on the head dress will also help keep you cool. I have had to resort to this many times! Pull your head dress across your face as well to protect it from the sun. A damp head dress will also make the air you breathe in less dry, and so keep you hydrated longer.

Sandals are less effective in the desert because they expose the tops of your feet. If they're all you have, wrap something around your feet to protect them.

Also, be aware of the common blister, which in the desert, with its rough, abrasive sand, can happen much faster than you might be used to. They can be debilitating to your progress and you need to act early to prevent them (see pages 397–401). Thin under-socks work well, as does regularly emptying sand out of your socks, boots or shoes.

IMPROVISED DESERT SHELTERS

When people think of deserts, they often imagine dunes, *Lawrence of Arabia*-style. Some desert terrain *is* like that, in which case you'll have to rig up a shelter out of a coat, a blanket, a sleeping bag or a tarp – get *anything* up that will protect you from the sun.

If you're in more level, firmer desert terrain, and you have a tarp or a shelter sheet, it's worth knowing how to make at least one of two scrape shelters. Try to build them in the early morning or the late evening. They can be strenuous work, and you need to conserve your energy and stop yourself sweating.

The first scrape shelter assumes you have no cordage. Dig down into the sand. Spread your sheet over the hollow and place rocks around the edge to keep it in place. If you have a second sheet, you can reduce the temperature in the shelter by making a double layer, with a gap of about 60cm between them.

If you have cordage, you can tie your sheet to four stakes – or improvise, using your pack or a pile of stones. Again, a double layer is best if you can manage it. This shelter will also work without the trench.

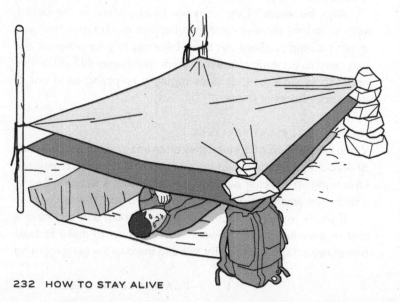

In less sparse desert environments, you have more options.

VEGETATION
Trees and bushes can provide effective shade. If you can find the right materials, you may even be able to make one of the shelters on pages 47–50.

CAVES AND OVERHANGS
Potentially a good solution. But remember: you're not the only thing in the desert that needs shade. Snakes, spiders, scorpions, mountain lions – all sorts of creatures could be using these desert features to survive.

VEHICLES AND WRECKAGE
Don't try and shelter in metal vehicles or aircraft wreckage: they'll turn into ovens and you'll quickly overheat. Instead, use the shade that they cast, or use them to support sheltering material.

WATER

Finding shelter doesn't just protect you from sunstroke and sunburn. It reduces the amount of water your body loses. But you're still going to have to find more to drink if you want any chance of long-term survival. Military personnel on desert ops will carry six bottles of water with them, which they know they'll need to have even the remotest chance of functioning at a normal level.

Solar stills, above ground and below ground, do work. See page 28 for the low-down, and fill it with cacti or other moisture-producing plants. Or urinate into the still.

It's also possible to gather dew and condensation by spreading out tarps, sheets of plastic or any bits of metal you have. Do this overnight, and lick off any dew you find in the early morning.

Alternatively, you can make a dew trap. Dig a pit about half a metre deep and line it with any material you have that will stop water leaching into the soil – plastic sheeting or even foliage. Fill this with stones. Condensation will collect on them overnight, which you can lick off before sunrise.

Animals need water just as much as you do. So the following are often indications that a water source is nearby:

- Animal trails. If they intersect, that's an especially good sign that more than one species is heading for water.
- Animal droppings.
- Bird flight. But not carnivorous birds, like vultures, which get their fluids from the carrion they eat.

If you find a dried-out river bed, you can try to dig a water hole. Again, don't do this in the heat of the day – you'll lose loads of fluid in sweat, and might not even find any to replace it. The lowest point of the outside bend of the river is the best place to dig.

Desert plants are getting their water from somewhere. Their roots will show you where to dig.

It's a bit of a myth that cacti will supply you with water. They are often poisonous and sometimes so disgusting that they make you vomit, which will dehydrate you further. But some cactus fruit, such as the prickly pear, can be mashed up and drinkable liquid squeezed out of it.

FOOD

The heat will sap your appetite and, as always, if you don't have enough water, you should avoid eating because it will dehydrate you further.

Food is hard to come by in the desert, but not impossible. The following are edible:

SNAKES

Best killed by pinning the head down with a stick and bashing it with a heavy rock. A high-risk way of getting supper, especially if you don't know if it's venomous. But all land snakes can be eaten. Gutted, skinned and cooked they taste pretty good. Raw, not so much – plus they then carry the risk of parasites and bacteria. But I have eaten many raw in extremis. You can also prepare the meat into thin strips that can be dried in the hot air and made into snake jerky. Nice!

SCORPIONS

In some parts of the world scorpions are eaten as a delicacy. To me, they always taste pretty grim. As with snakes, cooked scorpions taste a little better. If you cook them over a fire, the heat breaks down the proteins in the venom, making it edible. You *can* eat them raw, but you'll have to cut off the stinger at the end of the tail first. Remember: the bigger they are, the safer they are. The really deadly fat-tail or deathstalker scorpions have very small pincers. They don't need big ones – their punch is in the power of the venom in their tail stinger.

BIRDS

Birds often follow you in the desert. This is particularly true of carnivorous birds such as vultures, who are always on the look-out for carrion. If you lie still on the ground, using yourself as bait, they will start to come close. You may have a chance of grabbing one. Remember to lie in the shade, though.

Alternatively, you can use a bird trap. The following is an ancient device called an ojibwa bird pole. It works on the principle that if there is no natural perch nearby, a bird will settle on one that you provide. So, it is well suited for wide open spaces like deserts.

1. Find a straight pole between one and two metres high.
2. Sharpen one end with your knife. Drill a hole in the other end with the knife tip.
3. Tie a rock to a thin piece of cordage. The rock needs to be about the same weight as the bird you are trying to catch.
4. Thread the free end of the cordage through the hole and tie an overhand knot in it so that when the knot is by the hole, the rock is raised from the ground.
5. Insert a small stick, about 15cm long, into the hole, allowing the knot to hold it in place.
6. Make a noose in the loose end of the cordage, and drape it over the perch as shown.
7. When a bird lands on the perch, it will dislodge. The rock will fall, and the noose will tighten around the bird's feet.

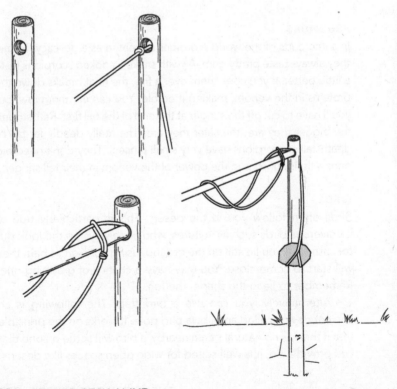

K.I.S.S.

(KEEP IT SIMPLE, STUPID!)

Shelter is your first priority. The sun can kill you in hours.

Water: avoid losing it, try to locate it.

Rest by day, move by night.

Only eat if you have enough water. Reptiles, insects and birds are your best bet.

HOW TO

SURVIVE

IN THE

SNOW

Sub-zero temperatures bring with them a whole range of survival problems. Moving effectively from A to B is one of them – see pages 249–54 for more on that. Your main survival priorities, however, remain as follows: protection, rescue, water, food. Or: Please Remember What's First.

- *Protection*. From the elements – just as in the desert the sun is your number one enemy, so in the snow the wind and cold will kill you fast if you ignore them. And protection from any dangers such as crevasses, avalanches, rivers, rockfalls and cliffs.
- *Rescue*. It is no good surviving for a few weeks in a snow hole, only for the rescue services never to be able to locate you. Make yourself visible with large SOS signs in the snow, made with branches for example. Or with an improvised flag pole made with a tarp and a branch. Make it easy for rescue services to pinpoint you from the air.
- *Water*. Keeping hydrated, whether you're in the desert, the jungle or the mountains, is vital. Think water before food. Act early to find it and purify it by boiling. This is the surest way to kill parasites and bacteria that if drunk will simply make you sick and dehydrate you faster.
- *Food*. Start thinking about and preparing to find food early on in your escape – see the section on tracking and trapping (pages 107–13). Remember, you can last three minutes without air, three days without water, three weeks without food. Act accordingly.

STAY BELOW THE TREE LINE

If you're in sub-zero mountainous regions, there comes an altitude where trees cease to thrive. Avoid going any higher than this. You have many more survival options below the tree line. You have wood for shelter and fire. You have plants and animals for nutrition. Above the tree line, you have none of this. Stay in forested areas whenever you can.

SHELTER

Finding some way of protecting yourself from exposure to the elements is essential. Don't underestimate the deadly effects of the wind. Wind chill factors massively increase heat loss. Nine times out of ten it's the wind that kills, and shelter is your only protection from this.

CLOTHES

Your clothes are your first – and best – defence against the elements. Get it right, and you can keep your core temperature above the critical 36°C mark. Get it wrong, your core temperature will drop below this level and you'll become hypothermic very fast. (See pages 442–3 for how to treat hypothermia and frostbite.)

Most survival experts will tell you that the best clothes for sub-zero temperatures are breathable Gore-Tex jackets and woollen socks, clothes and gloves. That's all true, but in a survival situation, the clothes you wear will be determined by what you have with you. To make them work hard, remember the acronym COLD.

- **C** It is important to try to keep your clothing *clean*. Not always easy in the wild. But if your clothes become dirty and greasy they lose some of their insulating qualities. Wash them regularly in streams or lakes if you can, but only if you have a fire to dry them.
- **O** Avoid *overheating*. When you get too hot, you sweat. Your clothes get wet, which means that later your body temperature will reduce further.

- **L** Wear your clothing *loose* and in *layers*. The air pockets in between the layers act as an insulator. Lots of loose layers are better than one thick jumper.
- **D** Keep your clothing *dry*. Wet clothes will sap the warmth from your body quicker than almost anything else. Wet socks especially can be a killer. In this environment, wet equals dead. Dry clothes by making a fire or wringing them out. Put socks in your armpits to warm and dry them during the day.

SUB-ZERO EXPEDIENT SHELTERS

A cave or other rock formation can be a life-saver. But again, remember you're not the only creature out there that needs shelter. First, examine the area for signs of animals – footprints, faecal matter, disturbed vegetation. You don't want an argument with a polar bear about who sleeps where.

And remember, too, that snow is a good insulator because it contains a high concentration of air – see pages 51–2 for information on how to build a snow hole. Here are three more ways of using snow's insulating qualities to keep you protected from the elements and your core temperature where it needs to be.

TREE SHELTER

Dig a pit around the trunk of a low-hanging tree. If you can line the base and walls of the pit with foliage, it will increase the insulating properties of your shelter.

TARP SHELTER

If you have a tarp or other blanket, you can cut blocks of ice to create a tarp shelter like this. Cover the sides of the shelter with foliage, or with more snow/ice. Position the shelter against the wind, so that it does not blow into the sides.

BLOCK SHELTER

If you have a snow saw and shovel, you can cut slabs of ice and use them to protect a snow trench as shown. Again, position this against the wind so it doesn't blow into either end.

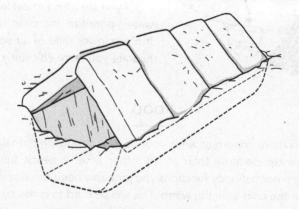

SNOW GLARE

The intense brightness of sun reflecting off snow can cause snow blindness. It's a bit like sunburn of the eyes: painful and debilitating. If you don't have snow goggles, you need to wrap cloth, or a strip of cardboard, round your eyes. Cut small slits in it to see through. Use charcoal from your fire to blacken the skin under your eyes (like American footballers do with eye block), which will reduce the glare.

WATER

If you're surrounded by snow and ice, you're surrounded by water. Moreover, snow and ice – as long as they're not obviously contaminated or taken from areas of standing water – tend to be very pure. Purify or boil your water if you can, but the risks associated with not doing so in this environment are fewer.

However, it's not as straightforward as that. Eating snow or ice can reduce your core temperature dramatically. If you can melt it, you must. You can do so using a fire, like this:

Alternatively, put ice or snow in your water container and hold it in your clothes where your body heat will melt it.

If you are going to eat frozen water, prioritize ice over snow. It has a lower ratio of air so will hydrate you more efficiently.

FOOD

Food is more important when your body is struggling to maintain its core temperature than at any other time. It needs fuel. To maintain normal body functions you probably need twice as much food in the cold as in the warm. This will be hard to come by.

In forested areas, you can look for vegetation and trap mammals (see pages 107–13). In Arctic conditions, your main food source will be fish. If you're by the coast, look for shellfish such as clams and mussels – they have a high calorific content. By frozen lakes or river beds, you need to learn how to ice-fish.

Use your knife to carve a hole in a frozen lake or river (see below for how to move safely across these areas). Attach a fishing line, hook and bait (see page 101 for how to improvise these) to a long pole as shown, and dangle your line into the water. You can stop the hole freezing over again by stuffing it with foliage or even – counter-intuitively – piling it with snow, which will insulate it. Remember: more ice holes means potentially more fish.

CROSSING FROZEN LAKES AND RIVERS

If you need to cross a frozen lake or river, be extremely careful. If you are at high altitude and the temperature has been sub-zero for some time, the ice might well be strong enough to take your weight. But don't be blasé. Many people die every year falling into frozen water. It can reduce your core temperature in minutes, and the water and semi-broken ice can be incredibly difficult to escape from. The freezing temperature rapidly robs you of the ability to operate your arms, hands or feet effectively, making any form of positive movement near impossible. Death is then all too common.

So, if you're crossing a frozen lake or river:

- Look for signs that the ice is thick enough to sustain your weight. There's an old saying: Thick and blue, tried and true; thin and crispy, way too risky. Strong ice tends to be a clear blue, free of snow and without any visible air bubbles. If it's white, opaque or grey it's likely to be weak. You should definitely avoid it if you see: water near the edges; water moving underneath the ice; cracks or holes; undulations on the ice caused by water currents or wind.
- Once you are confident that the ice can be crossed, remove the waistband and one strap of your pack. That way, if you go in, you can remove it rapidly and use it for buoyancy. Strapped to your back, it can turn you face down and drown you.
- Take small steps and carefully probe the ice ahead with a ski pole or long branch.
- If you see cracks or thin ice, lie down: you need to spread your weight over as wide an area as possible.
- If you go in, you need to move quickly: your body and your brain and physicality will begin to slow down in a matter of minutes, maybe less. Get your rucksack off and use it as a buoyancy aid. Turn back and aim to get out the same direction you went in. That is the only place where you know there is strong ice.
- If you have ski poles, hold them at the base and jab the sharp ends into the ice at the edge of the hole. Use them to drag yourself out, wriggling on your belly and kicking your legs to give yourself extra propulsion.
- Once you're out of the water, get to shore and get out of the wet clothes and into dry ones as quickly as possible.
- If you have no spare clothing, consider rolling in the

powder snow to absorb the water from your clothes. Wring out all your clothes and socks, and work fast.

- If you have spare clothing: strip down and roll in the snow naked to absorb the water from your skin before putting on your spare clothing.
- In either of the above scenarios, start a fire to dry your gear and warm yourself. If fire is impossible, get moving as soon as possible to regain body heat and circulation.
- You can also hang the clothing on some branches until the water freezes, then bang the clothing to shake out the ice. This is an effective way of drying clothing.

CREVASSES

A crevasse is a crack in a glacier. Sometimes you can see them, sometimes they're covered in snow. They can be 10 feet deep or 200 feet deep. They kill, regularly. I fell into a crevasse on Everest. If it wasn't for the quick thinking of my companions, and the length of rope attached to my body, I'd be a dead man.

You can look out over a massive stretch of snow-covered terrain and not know that it's heavily crevassed. And these crevasses play tricks. In the morning, when it's cold, that snow can be frozen solid. It's called an ice or snow bridge, and it allows you to walk over the crevasse without even knowing it's there. As the day wears on, and the sun hits the ice, the bridges tend to 'sag' as they warm up. That's when people die: they fall straight through the bridge.

If you're travelling with companions, rope up in groups of two to five with about four metres of rope between each person. As you move, keep the rope fairly tight between you. If one person falls, the others can pull him or her out. If you tie knots along the rope, they'll act as little anchors to add friction against the snow if someone falls.

As you move, probe the ground ahead with a ski pole or branch. Learn to recognize signs of hidden crevasses. Shallow depressions of sunken snow are a clear warning. It is easier to spot these at the start or end of the day when the sun is low in the sky and casts shadows on the depressions in the snow. At midday they are hard to spot, and the snow bridges will be getting weak. Move early in the morning if possible.

Use local knowledge. Glaciers grow and shrink, so the crevasses move around year on year. But locals will have an idea where this year's crevasses are located.

Keep your eyes peeled for cracks in the snow. You're not just looking for cracks straight ahead of you: follow the direction of all cracks to see where the crevasse might lead.

Sometimes you can see the sinister shadow of a crevasse beneath the snow. If it's narrow enough, you might be able to jump over it. But I wouldn't recommend crevasse jumping unless your life depends on it, and always rope up first.

It is possible to set up a self-belay system if you have rope and some stakes, so that you can make yourself safe as you cross crevasses. Once across you can pull the rope through, leaving only the stake behind. To do this, bang in an ice stake or screw a minimum of five metres back from the snow bridge over a crevasse. Tie yourself into a rope, then run it through a carabiner on the stake and back to your belay device. Pay it out as you cross. If you fall, your belay device holds you and you can lock off and use a jumar (a clamp attached to a fixed rope that tightens when weight is applied and relaxes when it is removed) to climb out. Once you're across, pull the rope through and continue – all you've lost is one ice stake or screw.

If you are on ice and don't have any crampons to provide traction and grip, consider using your socks over the outside of your boots. The wet wool quickly freezes, effectively giving you much more grip than you would otherwise have. I have often used this technique to travel across glaciers when without proper crampons.

K.I.S.S.
(KEEP IT SIMPLE, STUPID!)

Stay below the tree line for shelter, food and warmth.

C. O. L. D.

Be especially careful of frozen rivers and lakes. Spread your weight by lying down and crawling if necessary.

Crevasses are always there if you are on a glacier – you just can't always see them. Rope up.

HOW TO
MOVE
ACROSS
SNOW

Moving across deep snow is very difficult. You can easily sink up to your knees or even your hips and get stuck. Perhaps more importantly, moving through snow takes a huge amount of energy, which is one of your most precious resources in a survival situation. And that's before you even factor in the debilitating effects of extreme cold weather. It's essential that you find ways of moving quickly without exhausting yourself.

The problem is one of pressure. All your mass is concentrated on the relatively small area of your feet. To give yourself a fighting chance of moving across snow in a survival situation, you need to spread your weight as much as possible. In practice, this means using either skis or – easier to improvise – snow shoes.

SKIING

I love skiing. However, like most skiing enthusiasts, the lion's share of the skiing I do recreationally is 'Alpine' skiing – carving downhill on skis where your heel is fixed into the bindings. Amazing fun, with off-piste and jumps, but not a whole lot of use in a survival situation. We're not trying to be James Bond here. We're trying to get out alive.

The problem with Alpine skiing is it focuses principally on moving downhill. In an emergency, it's much more likely that you'll need to move from A to B quickly across, up *and* down snowy terrain. So you need to turn your attention to two other types of skiing: Langlauf (or cross-country) and Telemark. Langlauf skiing is great for going along the flat and up slight inclines. Telemark will get you up steeper inclines. The difference between these types of skiing and Alpine skiing is that the heel is not fixed to the ski.

This technique goes back hundreds, even thousands, of years, long before the advent of modern ski equipment. People would shape skis out of long, thin pieces of wood, then strap just their toes to the middle of the ski. If you've ever moved across snow in a fixed-heel ski, you'll know that it's difficult to move uphill. If your heel is free, undulating ground is much easier to negotiate. You can cover tens of miles relatively quickly. So if you want to learn to survival-ski, Langlauf and Telemark are the way to go.

In a survival situation, when you have to cross ground quickly, your biggest problem is going to be moving uphill. Here are two techniques for doing it.

SIDE-STEP

Nothing fancy about this. It's a beginner's move, but it does the job. Move your skis so they're perpendicular to the slope you want to climb. This will stop the skis' natural tendency to slide down. Move your upper ski up one step, then move your lower ski next to it. Keep going.

HERRINGBONE

More complicated than the side-step, but faster if you can get the hang of it. Stand so you're facing up the hill. Move one ski in front of you so it's pointing outward at a 30° angle. Shift your weight on to this ski. Now do the same with your other ski. Keep going.

Of course, moving uphill might be easier using a loose-heel technique, but it's still hard work. You can buy cross-country skis that have high-friction bases to stop you sliding around. In a survival situation, it's worth remembering the old huntsman's technique of strapping animal skins to the bottom of your skis. The fur grips the snow, giving you extra friction and stops the skis from sliding back downhill. Nowadays you can get synthetic ski skins that do a similar job. I have even peed on the base of a pair of skis before, in order to get traction. When the pee freezes, the ice creates friction with the snow and it 'sticks', giving traction for moving uphill.

IMPROVISED SNOW SHOES

In a survival situation, the chances are that you're not going to have skis, so you'll need to find another way to spread your weight. It's possible to buy dedicated snow shoes – plastic or aluminium fittings for your feet that look like short stubby skis. If you go back just fifty years, snow shoes were a lot less hi-tech and resembled tennis rackets. (I remember once taking Roger Federer out into the mountains and coming across an area of very deep snow. I actually tied tennis rackets to his feet and they did the job just fine!)

In a survival situation, you'll need to use the same principle, using whatever you have to hand. This means improvising with anything you have or can find to make your feet bigger. In a woodland area you can improvise little 'rafts' for your feet. The classic way of doing this is to make a tennis-racket-shaped snow shoe out of a green sapling by bending it back on itself and tying the two ends of the loop, then lashing branches horizontally across it and weaving a mesh between them. That's fine if you have time and just the right sapling to hand. In practice and under pressure, you don't need to be so precise. Grab whatever branches and twigs you can, then weave them – up and under – into a flat platform that you can tie to your feet.

Alternatively, you could use large, sturdy pieces of bracken to spread the weight around a little.

I have tried and tested all of these techniques over the years in survival situations, and they all work pretty well.

At a push, you can simply snap off a large, sturdy branch with plenty of foliage and strap that to your feet. It'll be heavy, so you'll need to weigh up the advantage you'll gain from spreading your weight against the disadvantage of expending extra energy. But that's survival – a constant risk-benefit analysis.

K.I.S.S.

(KEEP IT SIMPLE, STUPID!)

Moving across deep snow is exhausting
and you risk becoming stuck.

The secret lies in spreading your weight.

Loose-heel skiing techniques above fixed-heel skiing techniques.

Improvise weight-spreading platforms for your feet
to reduce the pressure you exert on the snow.

HOW TO

DRIVE

IN

SNOW

AND

ICE

Driving in snowy, icy conditions is treacherous. Here's what you need to know if you have to do it.

HOW TO FIT SNOW CHAINS

If you can get snow tyres fitted, do that. If not, snow chains are your friend if you're driving through snow and ice. (You can also get snow socks, but these are less durable.)

Lots of people make the mistake of trying to fit snow chains for the first time when they most need them. Trying to work out what to do with a tangled mess of chains in the middle of a blizzard probably isn't going to end well. You should learn how to do it *before* you need to do it.

Snow chains come in different sizes. Make sure you get the correct ones for your vehicle.

If you can, fit snow chains to all four wheels. If you only have one pair, front-wheel-drive vehicles and 4 × 4s need snow chains on the front wheels. Put them on the rear wheels for rear-wheel-drive.

To fit them:

1. Remove the chain from its packaging and make sure it's tangle free. Place the arch shape of the chain behind the wheel. The plastic cable should protrude from either side.

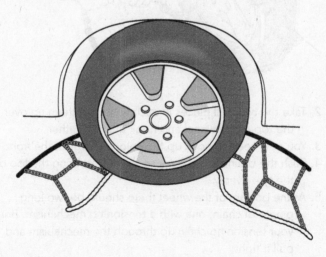

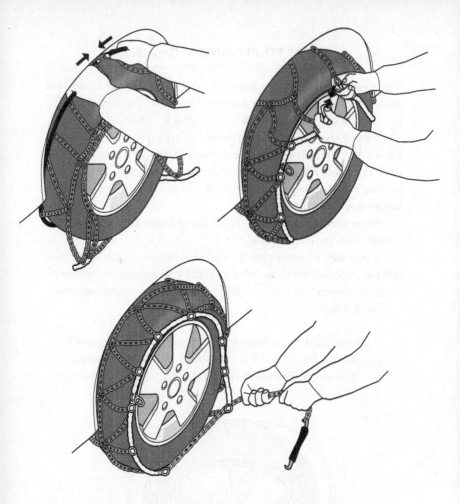

2. Take the two end pieces in each hand. Lift them up over the top of the wheel and hook them together.
3. You now need to hook up the coloured tags at the front.
4. Push the chains to the back so they're covering the top of the tyre evenly.
5. At the bottom of the wheel there should be two long pieces of chain, one with a tensioning mechanism. Bring your tensioning chain up through the mechanism and pull it tight.
6. Pull your tensioning chain up through the two loops on the chain facing you.

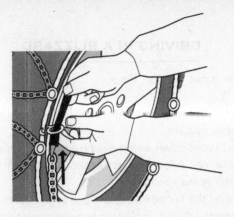

7. Now pull the elasticated end and hook it on to the top of the chain facing you.

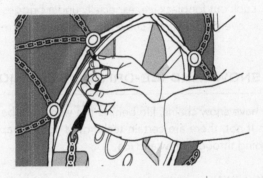

8. Your chains are now fitted, but they'll still be a bit loose. Drive your vehicle forward about ten metres, then re-tension them and hook the elasticated piece up as high as you can.

Don't travel faster than 30mph with snow chains on.

DRIVING IN A BLIZZARD

If you can sit it out, do that. If not:

- Clear all the snow from your vehicle's lights so you can be properly seen.
- Travel super-slowly and maintain extra-long stopping distances. Allow at least ten seconds to come to a halt, even at low speeds.
- Drive on the outside lane, or close to the centre of the road if it's a single carriageway: snow tends to accumulate and drift on the sides of the road.
- Drive in tyre tracks that have already been established.
- Try not to change lane.
- Look out for black ice, especially under bridges and overpasses where snow tends to melt then refreeze.

SNOW- AND ICE-DRIVING TECHNIQUES

If you have snow chains, life behind the wheel will be a whole lot easier. If not, there are certain techniques you can apply to keep you going through snow and ice.

PULLING AWAY
Pull away in the highest gear you can. This reduces the torque transmitted to the wheels, and so gives your vehicle more natural traction. If you keep stalling, however, your gear is too high.

If your wheels start to spin, release the accelerator until they grip again, then accelerate very gently.

DRIVING UP A HILL
This is tough in the snow. If you can go round a hill, do that. If you can't, you need to avoid using the accelerator too much as it will cause wheel spin.

Instead, try to build up speed as you approach the hill, so you have more momentum going up. Keep your foot very lightly on the accelerator.

If you're really stuck, and you're in a front-wheel-drive vehicle, try reversing up the hill. This shifts the weight of the vehicle on to the front tyres, where the power is.

DRIVING DOWNHILL

Do it slowly and in a very low gear. Try to avoid using the brake heavily. If you start to slide, you may have to release the brake completely while you regain control of the steering, then gently brake again.

CORNERING

Don't brake and turn at the same time. Slow down to the correct speed before you take a corner, preferably by using downward gear changes rather than heavy braking. Accelerate very gently as you emerge from the corner.

SKIDS

In general, if you start to skid you should avoid using the brakes as this makes the vehicle slide more.

Instead, take your foot off the accelerator and allow the speed to drop until you regain control. If you start to spin while skidding, you should normally steer into the direction of the spin. It feels counter-intuitive, but it's the best way to get the vehicle straightened up.

SNOW RECOVERY

If you're stuck on a bed of compacted snow, you need to clear away the snow from around the wheels. (Use a shovel if you have one, or your hands if you don't.) If you can, cut away a channel for the wheels to follow, which will help you gain traction. It's often easier to reverse back the way you came, because you'll have better traction on tracks you've already created.

If you're still stuck, try alternating between reverse gear and a suitable forward gear to get a back-and-forth rocking motion going. This can sometimes be enough to release the vehicle.

My friends and I were once stuck in the snow without snow

chains or recovery gear. But we did have string and extra T-shirts in the car. We used the string to wrap the T-shirts round the tyres. It didn't last very long, but it provided just enough friction to get us moving again and over the terrain that was causing us trouble.

Footwell mats are also very effective to lay in front of your wheels if you're stuck in the snow (or in loose sand or muddy terrain).

K.I.S.S.
(KEEP IT SIMPLE, STUPID!)

Practise fitting snow chains *before* you need them.

Travel slowly.

**Pull away in a high gear, but change to low gears
when you're at risk of slipping.**

**If you spin while skidding, steer into
the direction of the spin.**

HOW TO

SURVIVE

IN THE

JUNGLE

Much of the information in this book will stand you in good stead in the jungle. You'll know how to build a shelter, how to cross rivers and purify water, how to conserve energy, how to test if an obscure jungle plant is edible, how to avoid snakes and crocs, and how to light a fire even when it's damp. The jungle is truly the place where all your survival skills are required. That's why special forces soldiers head there to train.

You are surrounded by life in the hot, moist jungle, yet it is one of the most hostile environments for life there is.

GO WITH IT

Surviving in the jungle isn't just a physical battle, it's a mental battle. And really, you can't fight this environment. You have to become part of it. Understand that it can offer you everything you need to survive: shelter, water, food.

Also, understand this: you are *not* the strongest creature in the jungle, nor the fastest, nor the best camouflaged, nor the most venomous. But you do have the largest, most advanced brain. So slow down, calm down and use it.

LOOK UP

It might sound like simple advice, but you're far more likely to be killed in the jungle by a falling branch than by a venomous snake. Most animals know what their regular food source is, and stick to it. Inanimate objects are random hazards, but they fall from the high, thick canopy very regularly.

Be especially careful if curious monkeys are swinging overhead to take a look at you. Their movement through the trees will often dislodge dead branches. Being bashed on the head by one of those is a pretty uncool way to die!

A good friend in the SAS had both his shoulders dislocated by a falling branch in the jungle. He was lucky to live (see pages 436–7 for how to reset a dislocated shoulder in an emergency).

MAKE USE OF DAYLIGHT

If you've never been to the jungle, you'll be astonished how quickly it gets dark. And in the jungle, dark means dark. Don't put off your essential survival activities until it is too late or you will pay the consequences. Build your shelter or fire while there's still enough light.

BE WARY OF THE FOREST FLOOR

This is where you find snakes, leeches, scorpions and all sorts of critters. Always shake out your boots and clothes before putting them back on – you never know what will have taken up residence in them.

Always have your feet covered. If for some reason you've lost your shoes, improvise by wrapping them in cloth and/or tying tree bark around them for protection.

If you're making one of the shelters on pages 49–50, protect yourself from the forest floor by constructing a mattress of bamboo poles or branches with large palm leaves laid over the top.

PROTECT YOUR LARGEST ORGAN

Your skin is your largest organ. In the jungle it's vulnerable. The air teems with mosquitoes. The vegetation is thick with thorns. They can cause stings, cuts and welts. In the hot, humid environment these can become infected. You need to protect your skin at all costs. So:

- Cover it, even though you're hot and sweaty. Insects are attracted to salty sweat, which means they head straight for your wettest parts: armpits and groin. Wear long sleeves and make sure your legs and feet are covered. Tie a T-shirt over your head and let it hang down over the back of your neck.
- Use mosquito repellent if you have it. If not, rub mud into any exposed areas of skin. It might stink and feel uncomfortable, but as it dries it will form a protective crust to stop the mosquitoes getting to your skin. This has saved me many times. It's what animals do – and they always know best.
- Light smoky fires at dawn and dusk, when the mosquitoes are most active. The smoke will repel them. Burning dried dung or termite nests is especially effective at repelling mosquitoes.

- Move slowly and carefully through the foliage. The more you rush, the higher the chance of thorns and sharp branches tearing your clothes and skin.

LOOK THROUGH THE VEGETATION

When you first find yourself in thick jungle, the terrain can look the same in every direction. You have to learn not to look *at* it, but *through* it. Divide it into layers of varying depths. When you're looking in a particular direction, don't just look: really *see* your way through those layers. You'll be amazed, once you get your eye in, at the effect this has on your ability to navigate, and also to perceive potential threats.

GO ROUND OBSTACLES

Hacking your way through thick jungle is energy-draining work. Crossing swamps can be dangerous. If you can, go round them not through them.

LEARN TO LOVE INSECTS

There is plenty of vegetation in the jungle, but it's difficult to know which plants are toxic. Likewise, hunting meat in the jungle is hard. Most insects, however, are edible and a great source of energy-rich protein.

Look for insects in shady places – burrowed into the bark of trees, or in damp, dark holes.

You should avoid very brightly coloured insects as this can be a sign of toxicity – though it isn't always.

Mostly, insects are edible raw, but they can contain many parasites. Cook them if you can (fry, boil or grill).

Two of the best jungle insects:

- **Palm grubs.** These are found in the rotten bark of fallen palm trees. Cut them out of the wood with a knife and eat them raw or cooked. They can grow up to the size of a raw apple, and are easy to digest. From experience, look out for the pus as you bite into them. Edible, but tastes terrible!

- **Termites.** These are an awesome food source, and don't even taste too bad. If you come across a termite mound in the jungle, make use of it. Poke a knife or a wet stick into it and it'll come out covered with termites which you can safely eat raw (they are clean because they eat each other's faeces to restrict the number of parasites in their colony). They taste zingy, like citrus, which is the formic acid in them. It's antibacterial and harmless to you.

JUNGLE WATER

All the usual methods of collecting water on pages 25–31 work in the jungle, but here are two good additional ways for this particular terrain.

BAMBOO
Water collects in large, old bamboo poles. Give one a shake – if water's there, you'll hear it sloshing around. Drill a hole in the base of one of the sections, and one at the top to allow the air in, and the fluid will run freely out. Alternatively, remove the whole section and take it with you as a water-carrying device.

JUNGLE LEAVES
When it rains in the jungle, it really rains. Twist a large jungle leaf into a funnel shape to collect the rainwater into whatever containers you have. The more leaves you use, the more water you collect.

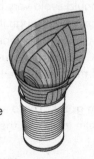

RIVERS LEAD TO CIVILIZATION

The most amazing story of jungle survival is that of Juliane Koepcke, who fell from an aircraft into the Peruvian jungle and lived. One of the reasons she survived that hostile environment is because she remembered her parents telling her one fact: rivers lead to civilization. If you're lost in the jungle, follow water. A tiny trickle will often lead to a small stream. A small stream will often lead to a larger one, and then to a river. Follow the river to find help.

See the next section for advice on crossing rivers, and pages 177–82 for instructions on making a raft – using the river's flow, as Juliane did, can be easier than walking alongside it. Just watch out for crocs or piranhas!

If possible, don't sleep by a river. There is a constant danger of flash flooding (see page 307). Better to get to high ground, and set up camp there.

K.I.S.S.
(KEEP IT SIMPLE, STUPID!)

Make the jungle your friend. Don't fight it. Go with it.

Stay covered and sleep off the floor.

Snack on insects and collect water when it rains.

Beware falling branches and rising water levels.

Rivers lead to civilization.

HOW TO
CROSS
A
RIVER

When it comes to extreme environments, people think of mountains, jungles, deserts and oceans. But make no mistake: rivers might not look it, but they are *dangerous*. They are incredibly unpredictable. Wherever there's a river, there's a source. It might be a mountain many miles away. When the snow on that mountain melts, or when there is heavy rain, all that water gets funnelled quickly down into the river. It's called flash flooding, and it can happen in minutes. I've seen rivers that look absolutely fine; a quarter of an hour later, they're wild and treacherous.

Rivers are also good at concealing their dangers. They can seem calm while hiding powerful undercurrents, eddies and holes. If you're going to risk crossing them, you need to know what you're doing.

ROPES

You might have seen pictures of people crossing rivers using ropes. Truth is, rivers and ropes don't really mix unless you're using purpose-made water-safety floating ropes. Otherwise they tend to sink and get wrapped around your body or other obstacles. Not good.

ENTRY AND EXIT POINTS

If you're going to cross a river, you need to choose these carefully.

ENTRY POINT

Unless you absolutely have to, you don't want to jump into a river from rocks, a bank or a cliff, because you don't know what you're jumping into. It could be unexpectedly shallow, or be hiding sharp rocks. You could easily break an ankle or a leg.

It's much better to wade in slowly. This means finding a river 'beach' where access to the water is easy. But there's a problem with this. If access to the water is easy for you, chances are that animals will come there to drink. Where animals come to drink, animals come to hunt. In extreme environments, where dangerous animals are commonplace, this means that the perfect crossing point could also be the most dangerous. There's not much you can do about it, however, except ensure that you have a heightened level of awareness. Look around carefully for any threats, and don't hang around. Once you have made your decision, get into the water quickly and go for it.

But above all with rivers, be aware of their hidden power. Too often I have seen people wade into a river to cross it and then unexpectedly get overpowered by the flow. Always give yourself an out, and plan carefully, even if this involves throwing a log in and seeing how fast it flows away and where it leads. Remember: once in it can be very hard to get out.

EXIT POINT

You need to be sure that, having crossed the river, you can get out. Again, this means avoiding rocks, banks or cliffs.

However, the river's current can move you downstream. Your exit point is unlikely to be directly opposite your entry point. You'll need to make a dynamic risk assessment of the state of the river in order to judge your exit point, and make sure you have line of sight towards it. Don't risk making the crossing if you don't know how you'll get out of the river.

RUCKSACKS

Before you enter the water, if you're wearing a rucksack you need to undo the waist strap and take one arm out. This is important even if you think the water's only knee deep. If you hit a hole in the river bed, a fully fitted rucksack can be so buoyant that it will force you face down into the water, or so heavy that it will drag you under. Half slung, you can (worst-case scenario) ditch it or (much better) use it as a buoyancy aid. Most rucksacks have waterproof liners, and if you've read the opening section you'll hopefully have your gear stashed in a bin bag or similar (see page 15). Either way, there will be air in the rucksack and it should float.

CROSSING TECHNIQUE

If the water is shallow enough for you to keep your feet on the river bed, your best crossing technique is to face upstream, brace yourself against the current and side-step across it. This means you're using the big muscles at the top of your legs to counteract the river's tendency to take your legs away. Use small steps, so you can feel for any changes in the river-bed terrain. And don't ever cross your feet over, as this reduces your stability and makes you liable to trip.

If you can, find yourself a sturdy branch, long enough to hold at a 45° angle with one end over your shoulder and the other end on the ground. This will give you an extra brace. I've done this many times. I can't promise that you won't get moved downstream (it's often happened to me), but it will give you significant extra stability.

FAST-MOVING CURRENTS

As I said, be aware of a river's hidden power. Think rapids and white water. You also need to be aware of forest 'furniture' – branches, tree trunks and the like coming downstream. If a heavy branch hits you at speed in the head, it can easily concuss you and then you will drown.

SLOW-MOVING CURRENTS

These can still be dangerous. There can be hidden threats – rocks, animals and the like. If the water's shallow, you can probably wade across. If it's deep, things are trickier, especially if you've got a pack. How are you going to give yourself more of a chance if you've got to make that hero-swim?

Your best bet is to get naked. Check out this picture of one of my climbing heroes, George Mallory, standing near Everest just after doing a river crossing.

Mallory wasn't getting his kit off just for the fun of it. You can't swim effectively with lots of heavy clothes, and you won't need your boots for stability. Put your clothes and boots in your rucksack and seal it up as much as you can. If you have a big bivi sheet or an old tarp, wrap it up in that. You can now use it as a buoyancy aid, and you'll have dry clothes on the other side. (Of course if you are wading through, keep your boots on for better grip and to help cope with any sharp rocks in the river.)

WHAT TO DO IF YOU FALL
INTO A FAST-MOVING RIVER

You've fallen in. You've lost your footing. You're moving downstream fast. Your best bet is to position yourself on to your back, your feet facing downstream, with your hands behind your head, ready to fend off any rocks in the river, or to bat out of the way any loose debris. The river will take you the route of the main flow. This can be bad news.

Be on the look-out for eddies, where the flow slows and doubles back on itself as it passes an obstacle. These areas of slower-moving water to the sides are life-savers. When you see one, swim like hell for it. You might have to 'punch' through the eddy-line, where the two different flow-directions meet. At this point, give a surge of power to cross into the slower water.

K.I.S.S.
(KEEP IT SIMPLE, STUPID!)

Rivers are dangerous: they can turn on you in an instant.

Choose entry and exit points carefully.

Fast-flowing shallow water: face upstream and
use a branch to brace yourself.

Slow-flowing deep water: remove your clothes
and use your rucksack as a buoyancy aid.

HOW TO

SURVIVE

UNDERGROUND:

CAVES,

TUNNELS AND

SEWERS

Caves can seem like a good idea when you're looking for shelter. Sometimes, when they're dry and vented, they are. But they can be dark, cold and damp. If you light a fire in one, the smoke can linger. They can be infested with bats, including vampire bats whose bite will make you bleed and bleed. And they can be full of guano or bat poo, which stinks and often carries disease. And, of course, what looks like a good shelter to you might well look like a good shelter to other dangerous animals, from snakes, spiders and scorpions to larger beasts such as mountain lions.

Caves can also be the entrance to a very hostile underground environment. The Mayans used to think that caves led the way to the underworld. They called them Xibalba, the place of fear. Having fought my way through many water-filled caves, I can tell you that's not a bad name. And many of the problems you're likely to face in a cave system also hold for tunnels and sewers. If you find yourself in one of these environments, you're going to be presented with a whole range of difficulties.

Be aware that flash flooding is a big risk underground. It can happen in literally seconds, so this is not an environment where you want to linger. You need to get out and get safe as quickly as possible. Easier said than done. You have to battle with the darkness and, crucially, the cold: hypothermia is one of the biggest potential hazards that you'll face. Your chances of survival in such an environment are precarious.

DARKNESS

Our eyes are quite good at adapting to the dark. Outside, even on the darkest nights and in the thickest forests, we'll gain some night vision after about ten minutes, even more after forty. Underground, however, this often isn't the case. If there's no ambient light for our eyes to use, we're completely blind.

If you're moving in such an environment, therefore, you need to be careful. If the ground is dry, consider crawling. This means you'll feel any sudden holes or cliffs in the terrain *before* you step into them. Be aware, though, that body contact with the cold rock can quickly sap your warmth.

If you're feeling the way ahead with your hands, have the *backs* of your hands facing forward, rather than your palms. Rock walls can be sharp and there can be all sorts of hidden dangers. You don't really want *any* part of your hand to be damaged, but it's best to protect your palms.

Underground, light is life. If you have a head torch, or other light source, use it sparingly: its battery is finite. Certainly only use one at a time, and if you have different power outputs, switch them to the dimmest setting. If you need to light a flame to see, be aware that there is a risk of naturally occurring flammable gases underground.

COLD

If you have any extra clothing in your pack, get it on quick. Underground environments can sap you of your body heat quicker than you know. If you can wrap something – anything – round your head to prevent heat loss, do so.

Keep moving on the spot if you feel yourself getting too cold, but not so much that you sweat as this will bring your body temperature down further.

If you have other people with you, huddle together for warmth.

If you need to get into water – and there's a good chance that you will (see below) – your best bet is to get your clothes off and store them in a waterproof bag in your pack, if possible. That way you'll have dry clothes to put on when you emerge.

If you've nothing in which to carry your clothes and you have to swim in them, take them off once you've emerged from the water and wring them out thoroughly before putting them back on. They'll sap your warmth less that way.

FINDING YOUR WAY

If people know that you're in a cave or tunnel they are likely to send help if you don't return, so your best bet is to stay put. But that's not always the case. If you're going to move through a tunnel, cave or sewer system, here's what you need to know.

Underground systems can be maze-like. To avoid getting completely disorientated, you need to be able to retrace your steps. So:

1. Try to leave markers, such as piles of stones, as you go along. Make the shapes you build unnatural so you can't mistake them for natural formations.
2. Remember that caves can look completely different when you retrace your steps. Look back often, so the cave's layout becomes familiar from a different aspect. Make a mental note of any features along the way. I call this mental mapping.

Remember, too, that air will get in by any means possible. If you feel a draught, follow it: it will likely lead to an exit.

And if you find flowing water, it's going somewhere. If you follow it, there's a chance that you'll find the place where it leaves the cave system. This may not always work for you as an exit point, but you might luck out. Be aware of water that runs into tunnels with no way out. Caution, caution, caution at every turn. Always make sure you have an 'out' to retrace your steps.

If you're in a group, hold hands as you move – it's very easy in the darkness for someone to be left behind.

Ration your food and water very carefully. Cave and tunnel systems can be massive. There's a chance you're going to be there for a long time.

K.I.S.S.

(KEEP IT SIMPLE, STUPID!)

Underground, light is life: preserve your light source.

In the darkness, feel your way slowly, palms inward.

Hypothermia is one of your biggest enemies.
Do everything you can to keep warm.

Make markers as you advance.

Follow draughts and flowing water. They're
getting in and out from somewhere.

LIFE-
OR-
DEATH
SITUATIONS

HOW TO

SURVIVE

AN

EARTHQUAKE

There's no way of accurately telling when an earthquake will happen. But when one hits, it can kill tens, even hundreds, of thousands of people.

Although minor tremors can occur anywhere, major earthquakes happen along natural fault lines in the Earth's crust. This means that, although it's not possible to predict *when* an earthquake will happen, seismologists do know *where* they are most likely to happen. If you're in an earthquake-prone area, it's essential to have an action plan for when the worst happens.

WHAT DOES AN EARTHQUAKE FEEL LIKE?

This depends on how strong the earthquake is, and how far from the epicentre you are. In general, however, being at the epicentre of a major earthquake feels like a sudden jolt, followed by anything from a few seconds to several minutes of very violent shaking. It'll be difficult to stand up.

WHAT TO DO

This depends solely on where you are when the earthquake happens.

IF YOU'RE INDOORS
Stay there. This might not feel like the right thing to do, but you're much more likely to come to harm if you try to leave the building once the earthquake has started because there will be so much falling debris – not to mention that it'll be very hard to walk.

Instead, position yourself so that if there *is* falling debris it's not going to hit you. So: the advice is to *drop*, find *cover* and *hold on*.

- **Drop.** Get on to all fours – it's easier to move around like this when the building is shaking, and by bending over you protect your vital organs.
- **Cover.** You should immediately cover your head and neck to protect yourself from falling debris. If you're at particular risk from falling objects, crawl under something sturdy like a table. If there's no such shelter, crawl towards interior walls – keep away from exterior walls (the most dangerous part of a building during an earthquake) and windows. Get away from tall furniture against the wall and from objects hanging on it, too.
- **Hold on.** If you're under a shelter such as a table, hold on to it so that you can move with it if it shifts. If there's no shelter, hold on to your head and neck until the shaking stops.

In high-rise buildings, expect the sprinkler systems and fire alarms to go off. Drop, cover and hold on, and do *not* use the lifts.

If you're in bed, you should try either to get under it or pull the mattress over yourself. Get on to your front to protect your vital organs, and cover your head and neck with your arms.

IF YOU'RE OUTDOORS

If you can safely do so, get to a clear area away from tall buildings, power lines and trees. Drop, cover and hold on: even if there's nothing above you, you can be hit by debris from the side during an earthquake.

If you're on a hillside, get to the top as quickly as possible. Earthquakes can cause landslides, and you'll never survive getting caught up in a large amount of fast-moving earth and rock.

If you're by the sea, drop, cover and hold on. Beaches are fairly safe initially, but earthquakes can cause tsunamis (see page 310). You need to get to high ground as quickly as possible.

If you're below a dam, you need to be prepared for catastrophic flash flooding if the dam breaks (see pages 306–9). Get to high ground at the side of the valley.

IF YOU'RE IN A VEHICLE

If you're driving, pull over immediately. Avoid bridges, power lines, signs and tall buildings. Put the handbrake on and stay inside the vehicle. If there's room to crouch down below the level of the seats, do so: you'll be a little more protected if anything falls on the vehicle. If a power line falls on the vehicle, wait there until a first responder brings help. (The danger comes when leaving the vehicle, stepping down and creating a 'bridge' for the electricity through your body that will easily kill you. If you have to exit the vehicle or no assistance is coming, leap as far as you can from the vehicle so that no point of you is connected simultaneously between car and ground.)

AFTER THE QUAKE

Once the tremors stop, you need to leave the building or any damaged outdoor area and get to an open space.

This won't necessarily be possible. If you're trapped in a building with no clear path to safety:

- Cover your nose and mouth with a piece of your clothing: there will be a lot of dangerous particles in the air and you don't want them in your lungs.
- If you have a phone, use it – and be aware that in the aftermath of an emergency text messages might get through more easily than voice calls.
- Tap on pipes or walls to help the emergency services locate you – use the SOS signal (see page 87).
- Don't try to move about unless your life is threatened: you might dislodge more debris and make matters worse for yourself.

If you're outside, don't take shelter in damaged buildings: they're now unsound, high-risk locations. Build a shelter out of debris instead.

Be aware that water sources might be contaminated because of damage to sewer systems.

Avoid making sparks, lighting matches or using lighters – earthquakes cause gas leaks.

And remember: earthquakes often have aftershocks. They might be less violent, but they're still dangerous. Be prepared for them. During aftershocks, all the advice above holds.

K.I.S.S.
(KEEP IT SIMPLE, STUPID!)

Drop, cover, hold on.

If you're on the slope of a hill, get to the top.

If you're in a vehicle, stay there.

Anticipate aftershocks.

HOW TO

SURVIVE A

VOLCANIC

ERUPTION

Areas prone to volcanic eruption are often the same as those prone to earthquakes – they occur where the Earth's tectonic plates meet. I've been to numerous active volcano sites around the world, and they are awe-inspiring reminders of the power of Mother Earth.

There are three categories of volcano:

1. *Extinct* – a volcano that is not expected to erupt again.
2. *Dormant* – a volcano that has erupted in the past two thousand years, but not recently.
3. *Active* – a volcano that has erupted recently, or is expected to erupt again.

It can be difficult to know if a volcano is dormant or extinct, but both are safe. Active volcanoes kill. In 1985, for example, the eruption of Nevado del Ruiz in Colombia killed more than twenty-three thousand people.

VOLCANIC HAZARDS

When people think of volcanoes, they think of lava. Lava flows are volcanic hazards, but they're not the only hazard nor even the most dangerous. Here's what you need to look out for in the vicinity of active volcanoes.

LAVA

This is molten rock flowing out of a volcano. The hotter it is, the runnier it is and the faster it moves. Most lava flows, however, are quite sticky and don't move much faster than a person on foot.

That doesn't mean they're not dangerous. Lava can be hotter than 1000°C, and will burn vegetation, buildings and humans alike.

ASH

Volcanic ash clouds famously cause flights to be grounded. The fine ash, which is actually powdered rock, can take out ground transportation, collapse buildings, cause power and comms outages and smother crops.

PROJECTILES

A volcanic explosion can spew fragments of rock high into the air, which can land anywhere over a massive area around the volcano. If these projectiles hit you on the head, they'll probably kill you.

LAHARS

These are mudflows formed from a mixture of water and volcanic debris. They can travel at up to 120mph and have the consistency of wet concrete. Lahars can form even in the absence of a volcanic eruption, or long after a volcano has erupted. They cause massive damage to anything they hit.

PYROCLASTIC FLOWS

These are heavy, fast-moving, devastating mixtures of lava, crushed rocks, ash and poisonous gases. They surge down volcanic slopes at hundreds of miles an hour and can reach temperatures around 700°C. They destroy everything in their path. Big danger.

GAS

Again, big danger. Volcanic gas clouds are not the most dramatic parts of an explosion, but they can be the most deadly, containing certain chemicals in concentrations that are potentially toxic to humans. These include:

- *Carbon dioxide*. Not specifically toxic to humans, but it's heavier than air. This means that pockets of it can collect in low-lying areas, replacing the air and causing suffocation in humans and animals.
- *Sulphur dioxide*. This can combine with water vapour (also emitted from volcanoes) to make sulphuric acid.
- *Hydrogen sulphide*. Extremely toxic in small amounts. Causes pulmonary oedema, unconsciousness and death.

HOW TO SURVIVE AN ERUPTION

If you're in a volcanic region, you're likely to get warnings from local news outlets that an eruption may be imminent. If you're instructed to evacuate, do so immediately – by vehicle if possible. If not, stay indoors. Shut all the doors and windows. Turn off any heating and air-con.

If there is very heavy ash fall, it can collapse the roofs of buildings. If you're told to leave the building for this reason, you need to protect your eyes and your respiratory system. Wear goggles or a diving mask. Ideally you'd have a respirator of some kind. If not, improvise with wide strips of damp cloth or torn clothing.

If you're off grid in the vicinity of a volcano, there are signs that an eruption may be imminent. These are:

- bursts of steam and gas from the crater
- the smell of sulphur
- earth tremors

If you experience any of these, evacuate the area immediately.

If you're caught in the vicinity of an eruption, your immediate priority is to protect yourself from falling projectiles. Look up. Protect your head and neck if there appears to be risk of these.

Your next survival priority is avoiding lahars and pyroclastic flows. They travel downhill, so you need to avoid low-lying areas. Try to put as much distance between yourself and the volcano as you can – ideally you want to be 10 kilometres away – but if this isn't possible, try to get to high ground. If there is a ridge line, shelter on the side furthest from the volcano.

Ash and gas are a constant threat, and both can kill very fast. If you can shelter from the ash, do so. Don't drink any water that has ash in it, and breathe through damp cloth or torn clothing.

K.I.S.S.
(KEEP IT SIMPLE, STUPID!)

Don't focus only on the lava. Volcanoes have many other ways of killing you.

Evacuate if you're told to and can.

Off grid? Get to high ground if you can't evacuate.

Protect your eyes and respiratory system.

HOW TO

SURVIVE

AN

AVALANCHE

Avalanches are common on snow-covered slopes, and so are avalanche deaths – even on relatively well-populated slopes. To increase your chances of survival in an avalanche situation, you need to learn how to recognize avalanche-prone slopes, know how to cross them if you've no other choice, and know what to do if you find yourself caught up in a snow-slide.

AVALANCHE-PRONE SLOPES

A slope is more likely to avalanche if:

- it has a steepness of between 30° and 45° (that's a gradient of 60–100 per cent)
- it has a convex profile – i.e. it's bulging outward (concave slopes are safer)
- there are no trees – these tend to anchor the snow in place
- it's north-facing in the middle of winter, or south-facing in the spring and summer
- there has been rain, or an increase in temperature immediately after a snowfall
- it's a leeward slope, where you are more likely to get drifted snow from the windward side

However, you should remember that all slopes have the potential for avalanche. You should take into account past and recent history and recent weather conditions, especially snow, rain, thaw and freeze. Speak to local guides if you can.

ESSENTIAL AVALANCHE SAFETY GEAR

If you know you're going to be in an avalanche area with other people, there are three essential items you should take:

1. *An avalanche transceiver.* This is an electronic beacon that you keep in your backpack. If you get caught in an avalanche, its signal will penetrate the snow above you. If one of your companions is lost, you set your own transceiver to search, and it will direct you towards their signal.
2. *A probe.* This is an extendable pole that you use to penetrate the snow in an area indicated by your transceiver. If someone's under the snow, time is short: you need to know you're digging in the right place.
3. *An avalanche shovel,* to help you get into the snow quickly.

SNOWPACK TESTS

Snow builds up in layers. The first snow of the year will freeze, then another layer of snow will fall and freeze. The process continues. These layers are not all equal. Some are harder and more compacted. Some are powdery and soft. An avalanche is more likely when a weak layer exists beneath a strong layer.

You can check for this by doing a snowpack test. Find an area of snow near to the one you want to cross. It should have the same aspect, elevation and snow depth. Use your knife, an ice axe or a shovel to cut out a block of the snow and examine the layers.

Warning signs in the snowpack are:

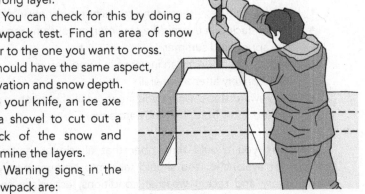

- harder, stronger snow immediately above softer, weaker snow
- very weak, sugary snow low down in the snowpack
- cracks in a weak layer of snow

Be warned: this is not an exact science. But it will help you decide how much risk you're willing to take.

CROSSING AN AVALANCHE-PRONE SLOPE

This should only be an option if you've decided that this is a risk you have to take in a survival situation.

First, put your rucksack over one shoulder, so that you can discard it quickly if you need to.

Next, understand that although avalanches do happen spontaneously, 90 per cent of people who die in an avalanche cause it themselves. So tread carefully and gently. Naturally you want to cross danger as quickly as you can, and you should certainly move as fast as possible. Do it lightly, though.

If you're in a group, rope yourselves together if possible, keeping about 15 metres between each person. Cross the danger area individually, while the remainder of the group try to anchor themselves to a fixed point.

HOW TO ESCAPE AN
AVALANCHE AS IT STARTS

In general, avalanches start in slow motion. It might be thundering and fast by the time it reaches the bottom of a slope, but the ground will move extremely slowly to start with. If this happens, you need to act fast. You have a matter of seconds to get above the line where the avalanche starts.

If you're near a ridge, quickly get to the other side of that ridge.

If this isn't possible, and you're caught up in the beginnings of the avalanche, your next survival priority is to get to its edges before it starts to gather real speed. If you can, head for hard cover like a big rock. This will act as protection from the momentum of the avalanche above you.

If you're fully caught inside the avalanche and it's tumbling you down the mountain, you've got real problems. You need to try to keep your head above the snow, and to 'swim' – preferably on your back so your face is pointing upward – on top of the snow and up the slope in an effort to resist the momentum. Easier said than done, but it can work for a small avalanche.

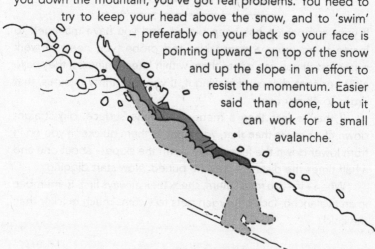

In all probability, you're going to be taken up inside the avalanche – it'll be like being spun inside a washing machine – and dumped violently at the bottom of the slope under several metres of snow.

On the one hand, you're in luck – it means you haven't been killed by the trauma of the slide.

On the other hand, you're in trouble. Snow is a good insulator, so nobody will hear you scream. It will be pitch black. You'll be encased in dense snow and you'll have it in your mouth, eyes and ears. Your oxygen supplies will be low: asphyxiation is the main cause of death in these scenarios.

If you have a transceiver, there's a chance that help will be on its way. In any case, your best chance of survival is early on.

The snow will very quickly start to harden, so you have to work fast. Make an air pocket around your head to give yourself some room to breathe. Take in a big breath so that your chest expands and makes more of a cavity in the snow before it starts to harden.

You need to get to the surface quickly. Trouble is, you won't know which way is up. To find out: dribble or, if your mouth is too dry, pee. Gravity will do its work on the fluid, and you need to burrow in the opposite direction. If the snow above you isn't too deep, you have a chance of survival.

DIGGING SOMEONE OUT

If your companion has been buried, and you have managed to locate them using a transceiver and probe, you need to work super-fast. If you can get to them within fifteen minutes, they have a 90 per cent chance of making it. If you take thirty minutes, that drops to 45 per cent.

If they're less than a metre below the surface, dig straight down. If it's more than that, you'll get to them quicker if you go in from lower down the face. Move down the slope – about one and a half times the distance they are buried. Now start digging.

When – if – you reach them, check their airways first. Remember, in an avalanche, lack of oxygen gets to victims much quicker than the cold.

K.I.S.S.
(KEEP IT SIMPLE, STUPID!)

Learn to recognize avalanche-prone slopes.

Carry a probe, shovel and avalanche transceiver
if moving across dangerous ground.

Tread softly but swiftly.

Get above or to the sides of a nascent avalanche.

Dribble or pee will always run downward.

HOW TO

SURVIVE

FLASH FLOODS

AND

TSUNAMIS

Flash floods, whether in the wild or in urban environments, can occur in literally minutes. They can be caused by persistent rain, by the build-up of water in lakes and rivers, or by natural phenomena such as earthquakes and high winds. They are common occurrences, and they regularly kill.

RIVERS

Rivers, even dry ones, are the hub of most flash floods. Even small rivers in non-extreme environments are prone to sudden and dangerous flooding. There is a river in Brecon where we do the Bear Grylls Survival Academy course called the Lightning River: one minute it's safe to cross, the next minute it's white.

Just because you don't have rain where you are doesn't mean a flash flood can't occur. Snowfall or heavy rain 50 miles away can lead to a sudden flash flood hours later. And people are all too often caught unawares.

So, if you're in the vicinity of a river:

- Look at the banks. If the river's only waist deep, but the banks are three or four metres high, nature's telling you something.
- Look at the plains surrounding a river. You'll see where flooding has occurred just as you can see the tide line on a beach.
- Don't camp too close to a river. Lots of people get killed every year doing this, swept away in a subsequent flash flood.
- Be aware of the weather conditions over the past few days. Snow up in the hills at the river's source can mean a flash flood days later.
- If you're in a dry river bed or canyon, there's still a risk of flash flooding because water can travel very fast along hard-baked earth. Keep a look-out for emergency escape routes at all times. Be especially wary of slot canyons where there is no escape either for the water or for yourself.

IF YOU'RE IN A BUILDING

If the flood waters are rapidly rising, your chances of survival are greater if you stay inside. If you're going to do this, then you must:

- Kill the electrical power and the gas at the mains.
- Close all the doors and windows.
- Move to the highest possible level and stay there until the flooding stops.
- Take a ladder with you if you can, in case you need to escape through the attic and on to the roof.
- If the flood waters rise dramatically and you have to get on to the roof, tie yourself and your companions to the chimney stack to avoid being swept away. (If you don't have rope, use sheets and blankets knotted together.)

IF YOU'RE OUTDOORS

Wherever possible, you should avoid trying to cross flood waters. They move fast and are *always* much more powerful than you might imagine. Even in knee-deep water you run a high risk of being swept away. Moreover, flood water often contains fast-moving debris – cars, boulders, even trees. Sometimes these items are hidden below the surface, so you won't even see them before they hit you.

If you have no option but to cross a fast-moving river, see pages 271–6 for what to do.

If you have to cross a flooded road, use a long stick to probe the ground ahead for hidden deep patches.

Be particularly wary of downed power lines.

In general, your priority must be to head for high ground.

IF YOU'RE IN A VEHICLE

Don't make the very common mistake of trying to drive through water if it's likely to be higher than your exhaust pipe. The engine can flood and the powerful force of the flood currents can easily sweep a car away.

If you find yourself in this situation, however, you need to be prepared to escape a flooding vehicle.

- Open all the windows and unlock all the doors while you still have engine or battery power. If you have a sunroof, open it. In a submerged vehicle, the difference in pressure between the inside and the outside means it's practically impossible to open a door until the interior of the vehicle is completely filled with water. You need to prepare your escape route before that happens.
- Release your seat belts. They can jam when your vehicle is knocked around. It's a good idea to keep a pocket knife in your glove compartment to cut through the seat belt in the event that it jams and you can't release it.
- In the event that electric windows can't be opened, you'll need to smash them. Don't bother with the windscreen: it's too tough. Kick out the side windows if you can. You can buy special tools to break the side windows quickly, which can be worth keeping in the car, but they won't work underwater. I have used the pointed end of the head rest before to do this, in a car that was sinking fast!
- If there are children, get them out first, even if it means grabbing them from the back seat and manoeuvring them out of the front window. Once water starts rushing in, they won't be strong enough to force their way out.

DRINKING WATER

In urban environments, flood water gets into the sewers and causes them to overflow and their contents to spread. Surface water, ground water and water distribution systems can quickly become infected with water-borne pathogens, fuel and other toxins. Don't drink water unless you're very sure it's safe. Boil it if you can, and see page 29 for other methods of purification.

In a rural environment, the rules for purifying water before drinking it still apply. There's a high likelihood of animal corpses rotting upstream.

TSUNAMIS

The word 'tsunami' is Japanese for 'big wave'. They are one of the most destructive natural phenomena. They are generally caused by underwater earthquakes that shift part of the sea bed. This causes an unstoppable wall of water that is impossible to withstand. Your only defence is to escape it.

If you are in a tsunami area, make a priority of recceing the official evacuation routes, or work out how you can get to high ground.

There may be official tsunami warnings that danger is on its way. These will be on radio, TV, by text message or from door-to-door contact with emergency personnel. Don't ignore them. Do the smart thing and follow the evacuation instructions quickly.

If you are by the sea, there are two natural warning signs that a tsunami might be on its way:

- an earthquake-type ground shake lasting longer than twenty seconds
- the sea receding unexpectedly and further than usual

If there are no official tsunami evacuation routes, you need to get to high ground. Be prepared to travel on foot: roads and bridges may have been damaged.

If there's no high ground, aim to get at least two miles inland.

If neither of these is possible – either because there's no time or a panicked mass evacuation is in progress – get to the top of the highest building you can see.

Don't rely on trees or other vegetation to protect you. They'll be swept away.

Whatever you do, *don't* make a conscious decision to stick around to look at the wave. If you're close enough to see it, you're close enough to be killed by it. Too many victims have been lulled into a mesmerized haze, staring at the sea as the tide races out suddenly. This is then followed by the surge. If you are caught exposed in this barrage of water, you are going to be at the mercy of its power. And that rarely ends well.

This mesmerized staring is simply a subconscious reaction to confusion, panic and fear. Being deadened into inaction. Fight this, and remember: stay calm, think fast but clearly, and then take positive considered action to ensure your safety and that of your family.

K.I.S.S.
(KEEP IT SIMPLE, STUPID!)

Get high.

Don't try to cross the flood water.

Escape from vehicles *before* they are submerged.

Don't drink untreated water.

The only defence against a tsunami is to evacuate.

HOW TO

SURVIVE

HURRICANES,

TORNADOES

AND

LIGHTNING

You only have to watch TV once a year, and see the regular devastation caused by hurricanes on the east coast of the United States, to know that these wild storms can represent nature at its fiercest. The good news is that nowadays hurricanes can be predicted accurately. But that doesn't stop people dying. Here's how to make sure it isn't you or your family.

WHAT IS A HURRICANE?

A hurricane, also known as a cyclone or a typhoon, is a wind above Force 12 on the Beaufort scale. I think it's worth remembering this, because it's a reminder that *all* strong winds are potentially deadly: much of the information here holds in the event of other violent storms that regularly kill.

Hurricanes develop at sea where warm air creates circling winds that can increase in speed as the hurricane grows. They tend to slow down as they hit land masses, but they generally occur between June and November in the northern hemisphere and between November and April in the southern hemisphere.

HURRICANE WARNINGS

These are issued in hurricane-prone areas. Do not ignore them. Understand that hurricanes are erratic, so it may not be possible accurately to predict where they will make landfall, or how strong they'll be when they do. Don't be blasé: prepare for the worst.

If the instruction comes through to evacuate, you should heed it. This is particularly important if:

- You are accommodated in a temporary structure. Mobile homes, camper vans and the like can easily be ripped from the ground by a hurricane. They are no protection.
- You are in a high-rise building. The force of a hurricane is stronger the higher up you are.
- You are by the coast. Hurricanes cause immensely powerful wave surges and coastal flooding.
- You are next to a river or on a flood plain. Hurricanes often bring torrential rain that can cause sudden and devastating flash floods.

EFFECTIVE SHELTER IN
URBAN ENVIRONMENTS

If you are in an urban environment, you will likely know that a hurricane is imminent. If you are unable to evacuate, you need to take shelter in a solid building. You don't want to be outside, because flying debris can kill you in an instant.

If you can get to a designated hurricane shelter, do that. There can be safety in numbers.

If not, preparation is key. You need to select a safe place in your building where you can sit out the storm for up to seventy-two hours. If you're in a single-storey house, your safe place should be near the centre of the house, away from windows and exterior doors. In a two-storey house or larger, you should position yourself near the stairwell on the ground floor, again away from windows and doors. If you can make your safe place a bathroom, do so, so you can use the toilet and have a water source.

Prepare a survival pack. You will need non-perishable food and water. If your safe place is a bathroom, fill the bath with water – the hurricane might put the urban utilities out of order. Otherwise fill as many containers as you can with fresh water for drinking and sanitation purposes. Aim for a gallon per person per day for drinking water.

Other items you need in your survival pack are:

- a radio and batteries so you can stay up to date with hurricane warnings
- prescription medicines
- first aid kit
- sanitation supplies (if not in a bathroom): a plastic bucket, toilet paper, bleach and detergent, plastic bags
- water purification tablets – you may still have running water, but the hurricane may have rendered it undrinkable
- a torch and batteries
- a whistle to signal for help

Before the hurricane hits, secure any loose items in the vicinity of the building. Switch off the electricity at the mains. In hurricane-prone areas, many houses have hurricane shutters over the windows. Make sure they're down. If you can, brace exterior doors from the inside. Don't bother taping up windows: it's no protection against a hurricane.

When the hurricane hits, it can be tempting to emerge from your safe place to witness this amazing force of nature. That's how people get killed. Stay indoors.

Don't be fooled by the eye of the hurricane. This can cause up to an hour of calm weather before the storm hits again. The sun might even come out before the winds return (because it's a circular storm they'll be coming from a different direction). You should only emerge when the hurricane alerts on the radio tell you it's safe to do so. If you have no means of accessing a radio, wait at least an hour after the winds drop.

SURVIVING HURRICANES IN THE WILD

If you're out of comms and in the wild, you won't have the benefit of hurricane warnings. I was once caught in a hurricane in the jungle and it was pretty terrifying, with trees being felled around me like matchsticks. I took shelter in the entrance of a cave. I have experienced several other hurricanes first hand, once at sea and once in the mountains. All of these occasions have been incredibly humbling reminders of the power of nature, and were exhausting experiences to come through.

Choose your battlegrounds in the wild carefully, and if in doubt re-schedule your journey.

In such a situation, whether or not it's officially a hurricane doesn't really matter. Extreme storms in the wild are brutal and dangerous. You need to be on the look-out for them, and know how to shelter.

BAROMETER ACTIVITY
Many wristwatches have a barometer function. They are particularly helpful in mountainous regions, but wherever you are a sudden drop in pressure generally means bad weather is on the way.

CLOUD FORMATION

Cloud behaviour is a good indicator of an incoming storm. There are many different types of cloud, which behave in different ways and predict different types of weather. But in a nutshell: if you see flat clouds becoming lower, or puffy clouds becoming higher, be aware of incoming storms.

FINDING SHELTER

If the winds seem to be getting very bad, you need to find natural shelter. In practice that means:

- caves (see pages 277–8 for the dangers they might present)
- ditches
- the lee side of cliff faces or rocky outcrops.
- the lee side of a forested area with big trees (avoid smaller trees as hurricanes can easily rip them up; and, if possible, remove any dead or dying low-hanging branches)

If none of these features are available, you need to get flat on the ground on your front with your arms covering the back of your head.

TORNADOES

Tornadoes, or 'twisters', can be even more destructive than hurricanes. They are violent columns of rotating air that reach from a thunderstorm down to the ground.

Unlike hurricanes, you can normally see and hear a tornado coming. The twister itself is recognizable, but warning signs include:

- a dark sky with a green tinge
- large hailstones
- dark, low-lying, rotating clouds
- an enormous roaring sound, like a huge engine

Most of the advice for surviving hurricanes holds for tornadoes.

If you're going to try to get out of the area, move perpendicular to its direction of travel.

LIGHTNING

You tend not to get lightning during a hurricane, but in the wild it is a very real danger during other extreme storms. Our bodies have a high water content, which makes them good conductors. Thousands of people die from lightning strikes every year. Here are some tips for avoiding them:

1. Use the 30/30 rule. If you can't count to thirty between seeing lightning and hearing the next clap of thunder, get inside. Don't emerge for thirty minutes after hearing the last thunderclap.
2. If you can't get indoors, try to head for a valley or ravine. Lightning tends to strike high points. Avoid exposed hilltops.
3. Avoid any large body of water: it conducts electricity and lightning is attracted to it.
4. For the same reason, avoid tall objects, especially trees. You contain more water than a tree, so there is a risk of lightning jumps as the electricity tries to find the path of least resistance.
5. Avoid large metal objects, and get rid of any you have on your person – think ski poles or ice axes.
6. Cars are generally safe places in a lightning storm, even if they're hit.
7. If you feel a tingling all over your body, as if your hair is standing up on end, it's often a sign that you're about to be hit by lightning. Crouch down – don't lie down. Make yourself as small as possible. Cover your ears with your hands and put your feet together. If lightning strikes the

ground nearby, you're at risk of electric shock. This position encourages current flow from the ground to avoid your body as it finds the shortest path back to the ground.

If you're in a group, spread out. One lightning bolt can harm several of you if you're bundled up. Also, if you're spread out, you can help if one of your companions has been hit, or receive help if it's you.

Lightning strikes are not always immediately fatal, but they can cause your heartbeat and breathing to stop. You'll need to administer, or receive, CPR (see pages 417–21).

K.I.S.S.
(KEEP IT SIMPLE, STUPID!)

If possible, evacuate.

Shelter from hurricanes and tornadoes in solid, sturdy buildings.

Prepare a survival pack.

Learn how to predict storms in the wild and act accordingly. Leave the ego at home!

Use the 30/30 lightning rule.

HOW TO
SURVIVE
A
SANDSTORM

Desert regions are very prone to high winds because of the way air masses of different temperatures mix in the atmosphere. Combine these high winds with loose sand and you've got yourself a potentially killer sandstorm.

These sandstorms present their own unique challenges. It's a bit of a myth that you're likely to be buried in sand. Instead, the main dangers are suffocation and disorientation.

FIND SHELTER

It doesn't matter what – a large rock or a parked car will do. I once used a camel carcass! Position yourself on the leeward side of the shelter and hunker down. But don't spend a long time looking if you know a sandstorm is coming. Shelter can be difficult to find in the desert, and when a sandstorm is moving in fast, your time is probably better spent preparing yourself for when the storm hits.

PROTECT YOUR AIRWAYS

If you get particles of sand into your respiratory system, you risk causing it severe systemic damage. Ideally, if you're travelling to a desert environment you should have a mask with you. If you don't, you need to cover your mouth and nose with a scarf, a bandanna or any piece of material – a cotton T-shirt, for example, will act as a basic filter. It won't be perfect, but it'll be a hell of a lot better than the alternative.

COVER YOUR EYES AND BODY

Goggles are best for your eyes. Ordinary glasses provide some protection. If you've got nothing, you can improvise some sand goggles by getting a strip of paper or (better) cardboard and making a couple of eyeholes in it with a pin or a pencil tip. Fit the strip to your head using an elastic band or some gaffer tape or plaster from your survival pack.

Once your airways and eyes are protected, you need to cover as much of the rest of your body as you can. Sand particles can painfully flay skin. Do whatever you can to avoid it being exposed.

HIT THE GROUND

Never try to walk through a sandstorm. You'll get lost and you'll exhaust yourself – those things can be wild. Get your back to the wind and hunker down, always making sure that your airways are protected.

The lower you get to the ground, the better, for two reasons:

1. Sandstorms can throw large bits of debris up into the air. The lower you are, the smaller your chance of being hit by one. You should cover your head and neck with your arms.
2. The heavier sand particles are close to the ground, while the finer ones are higher up. It's the fine sand particles that do the most harm to your respiratory system.

Before you hit the ground, you should orientate yourself. Know which direction you were heading in relative to the way you're facing now as you get low. Don't move during the sandstorm. The scenery can look very different once a storm abates, but you'll know which direction to continue travelling in.

If you are in a group, link arms. Sandstorms can be scary and disorientating. You don't want anyone freaking out and getting lost.

Be aware that, counter-intuitively, sandstorms can cause spectacular lightning strikes. See pages 319–20 for how to protect yourself from lightning.

IF YOU'RE IN A VEHICLE

Get to the side of the road and park up safely. Kill the engine. Wind up the windows. Keep the doors shut. You won't be able to see a thing, so you certainly shouldn't drive. And make sure your headlights are off. If they're on, other people travelling (unwisely) from the opposite direction will mistake your position for the path the road takes. This might make them drive off the road and into danger.

K.I.S.S.
(KEEP IT SIMPLE, STUPID!)

Find shelter if possible.

Protect your airways, your eyes and your body.

Keep your sense of direction.

In a car, get to the side of the road and kill the headlights.

HOW TO
ESCAPE
FROM
QUICKSAND

You'll find quicksand anywhere that sand and water meet – near river banks, lakes, marshes and by the sea.

You can forget those bits in movies where people get sucked under the surface of a quicksand, never to be seen again. Contrary to what most people think, it's pretty much impossible to drown in quicksand.

Quicksand is a mixture of granular material – not just sand, but also silt, mud or clay – and water. That makes for a very dense material. Twice as dense as the human body, in fact. Because it's so dense, humans float in it. You're unlikely to sink further than your waist or, worst-case scenario, your chest.

But quicksand is definitely *not* safe and can kill you in other more unpleasant ways. It is a 'non-Newtonian' fluid. That means it appears to be solid until it's slightly disturbed, when it takes on the properties of a liquid. Disturb it more violently, however, and it becomes more viscous, more difficult to escape. That means that the more you fight it, the more it resists you. Each time you pull up your leg, you'll find that suction drags it back down again. You won't drown, but you will be stuck.

When that happens, it's not the quicksand itself that's going to kill you. It's the sun overhead or, if you're by the sea or a river, the tide coming in, or a flash flood.

So you need to get out of there, fast.

RECOGNIZING AND AVOIDING QUICKSAND

Three tips for keeping clear of quicksand:

1. Be on high alert when you're trekking through quicksand danger zones. That's river banks, lakes, swamps and marshes, tidal flats and any areas with underground springs (which can include desert areas). If you see quicksand warning signs, don't ignore them.
2. In these areas, pay close attention to the terrain. You're looking for water bubbling up from underground, or terrain that looks rippled.
3. In danger zones, grab a stout stick and prod the terrain ahead of you as you walk. If you find that the ground gives way, take another route.

WHAT NOT TO DO

If you're stuck in quicksand and you're with a group of people, it's tempting to reach out and grab a helping hand. Unless you have no other option, that's not a good move. Firstly, you risk dragging the other person into the quicksand, which will mean two of you in the same situation. Secondly, it's easy to dislocate a shoulder if your companion is pulling you one way and the quicksand is pulling you another.

HOW TO ESCAPE

If you have a partner with you, make sure they remain free from the quicksand and get them to find you a strong stick. You can use this laid flat on the quicksand to help manoeuvre your body and limbs on to the surface. If they have a rope, they should secure it to a tree or around themselves and sit securely in a body belay position, acting as an anchor (see below). Put one arm through the rope so that it is loose across your chest, and use this to pull yourself against.

If you're able to loosen your backpack, do so. Lay it in front of you to use as ballast. Try to wriggle on top of it to help spread your weight.

Make yourself big. Spread your arms out, doing everything you can to keep them above water. This makes it even more difficult for you to sink further – it's the same principle as spreading your weight on thin ice.

And keep calm. I know everyone says this in a survival situation, but it's doubly important in quicksand because the more you fight it, the harder it's going to be to escape. Control your breathing and keep your movements slow.

Gently wriggle your legs. This makes a small space between your legs and the sand for water to seep into. This will loosen the quicksand a little and give you the opportunity – slowly – to work your way to the surface. Take your time and be patient. It is multiple small, slow, controlled, correct movements that will eventually allow you to escape. Each time you move a foot up an inch or two, give it a few seconds for the space you've evacuated to fill with quicksand again. This can be a very slow process.

Now, manoeuvre your chest on to the surface of the quicksand, as if you're about to do the breaststroke. As you're spreading your weight, you shouldn't sink. Continue gradually easing your legs out. When they're free, you should be able to monkey crawl over the surface of the quicksand to its edge.

WHAT NOW?

Your problems aren't completely over when you've escaped the quicksand. Your feet, groin and armpits are going to be covered with thick, wet, abrasive mud and sand. If you try to walk anywhere like that, you'll be rubbed raw in under an hour. The grit will make the skin on these tender parts of the body prone to rashes, boils and infection – which can be just as much a killer as the quicksand itself.

Get your clothes off and do whatever you can to get rid of the majority of the sand. If you can wash them in a river, do so. If your quicksand was caused by a desert spring, and the sun is searing overhead, work quickly so you can get those clothes back on to protect you from the sun.

K.I.S.S.
(KEEP IT SIMPLE, STUPID!)

Don't let people pull you out until they are properly prepared.

Stay calm: panicked gestures make things worse.

Use small movements to free your legs,
then crawl across the surface.

Use your backpack in front of you, a rope or even a sturdy stick
to help spread your weight and pull yourself to the surface.

Wash the sand from around your groin, armpits
and feet before continuing on your way.

HOW TO
ESCAPE A
FOREST
FIRE

At the Bear Grylls Survival Academy, we teach people how to make fire. In a Californian forest, one of my instructors was demonstrating how difficult it is to use a method of firelighting called the bow and drill method. He had put the tools down and, as he was talking, the bow and drill – made from the dry, parched wood of the forest – ignited within five seconds.

There are parts of the world whose forests have barely seen rain in years. It takes almost nothing to set fire to them. Once that happens, and the fire starts to spread, you have one of the most terrifying and deadly phenomena you'll ever encounter. Even in the UK forest fires can and do happen.

DON'T BE THE ONE TO START IT

Forest fires are often caused by humans. If you're not careful, they might even be caused by you. It only takes an ember or a little spark.

So: if you're in an area with a campfire ban, do not ignore it. If you build a campfire in a wooded area, make sure it's small. Clear the ground down to bare earth and surround it with a wall of rocks. Burn timber only, rather than material with lots of foliage – you don't want burning leaves floating off into the dry forest.

Read page 45 to learn how to extinguish a campfire properly.

LOOK FOR THE SIGNS

All forests are potential danger areas for forest fires, but you need a heightened level of awareness when you're in a particularly dry, parched coniferous forest.

The first sign of a forest fire will likely be the exiting of animals. If you see more wildlife than you're expecting, and they're all moving in the same direction, that's telling you something. The animals will sense the fire before you. Do what they do, and get out of there.

You will most likely smell and hear the fire before you see it. If you do, you need to leave the area quickly, because once you *can* see the fire, your situation is much worse.

UNDERSTAND WHAT YOU'RE DEALING WITH

The heat of a fully raging forest fire is unimaginable. We're talking maybe a thousand degrees Centigrade. Heat radiates a long distance from the flames. You'll be burned to death long before the fire itself actually reaches you, and inhaling the hot air can melt your lungs in a single breath.

The smoke and toxic fumes – of which there will be an unbelievable quantity – are as deadly as the heat, and will be blown towards you by the wind.

A forest fire can move at up to 20mph. That's faster than anyone can run for a sustained period. You're going to have to use your brain as well as your legs.

WHERE TO RUN

The direction in which a fire spreads will be determined by two factors: the wind, and the availability of fuel.

Firstly, try to establish which way the wind is blowing. The smoke will be a good indicator. You want to move upwind – that is, *into* the wind – if it's blowing towards you and the fire. If it's blowing towards you *from* the fire, you need to run at right angles to the wind and try to move *around* the fire.

Resist the temptation to get to high ground: fire travels more quickly uphill. Move downhill if you can.

Your next priority is to get to a natural fire break. In managed forests, there are often areas where the trees have been cut down to stop the spread of wildfire. Otherwise, you can head for water – a river, or a large lake. Don't be tempted to hide in a small pond. As soon as the fire gets anywhere close, the heat will boil both the water and you. If you can cross a river or a lake, do so, but don't assume you're definitely safe when you get to the far side. The heat of a forest fire can radiate across substantial gaps. This means the fire can jump. Sometimes you can't get far enough away.

WHAT IF YOU CAN'T RUN?

The wind's blowing in the wrong direction. There are no fire breaks. The flames are approaching. What do you do?

You have two options. Both are extremely high risk, and only to be considered when the alternative is death.

1. *Start your own fire.* The aim is to burn away the fuel in a patch of terrain. When the wildfire reaches it, all the fuel has gone and it has nothing to burn. Your fire will burn in the same direction as the wildfire, giving you a

patch of safer ground to head into. The bigger the fire, the more safe ground you'll make for yourself. You just need to be very sure you've correctly worked out which way the wind is blowing.

2. Go to ground. This is last gasp survival. Find, or dig, a ditch. Get in. Cover yourself with earth. The fire will move over you and the heat will naturally radiate upward. It will still be unbelievably hot, of course, and you will have an extra problem: the fire will consume much of your available oxygen as it passes over you. But in a worst-case scenario, this might just keep you alive.

IF YOU'RE TRAPPED IN A VEHICLE

Vehicles are not good places to be when there is fire all around: think of how quickly a metal saucepan full of water boils on the stove. However, a vehicle will give you some protection from the heat if there is no alternative, and people *have* survived this situation. Car petrol tanks tend not to explode unless the fuel is mixed with oxygen. To give yourself the best chance:

- Roll up the windows.
- Close all the air vents.
- Leave the engine running, but set the air-conditioning to recirculate.
- Drive away from the fire if you can. If you can't, stay inside the vehicle. Your chances are better in there than outside.
- Be prepared for hot air currents to rock the car violently.
- Get as low as you can and cover yourself with coats or blankets.

K.I.S.S.
(KEEP IT SIMPLE, STUPID!)

If all wildlife is moving in the same direction,
it's running from something.

Run into the wind if it's heading towards the fire. Run at right
angles to the wind if it's heading from the fire.

Head for natural fire breaks.

Head downhill, *not* uphill.

HOW TO

ESCAPE

A BURNING

BUILDING

On pages 41–6 you learned how to extinguish out-of-control fires. Sometimes you can't. If that is because you find yourself in a burning building, you're in one of the most perilous survival situations imaginable. You need to think fast and smart if you're to have any chance of making it out alive.

You also need to move quickly. There's a reason we say something spreads like wildfire. Your time is extremely limited. This is not the moment for stealth. Every second counts to get out of there.

IT'S NOT THE FIRE THAT KILLS YOU

It's the smoke. Thick, choking, blinding, toxic. In general, fatalities in fires are dead from smoke inhalation long before the heat or the flames get to them. So:

1. Keep low. As close to the floor as possible. Crawl if necessary. Smoke tends to billow up, so the concentration of toxic fumes will be lower nearer the floor.
2. Get any type of material – a rag, an item of clothing – and put it over your mouth. If you can get it wet, so much the better. It will filter some of the toxic fumes.
3. Shut your eyes if the smoke is thick. It will be pitch black anyway because as soon as fabrics (in particular) start to burn, they pump out blinding smoke. Keep it out of your eyes.

BE WARY OF INTRODUCING OXYGEN

At some point you'll need to open a door or a window to get out. Until that point you need to be very careful about letting more oxygen into the building because this just makes the fire burn hotter. Keep vents and small windows closed. If you see smoke around the edges of a closed door, or if it feels warm, don't open it unless you have to: modern fire doors can resist flames for thirty minutes. If you open it, you'll risk making the fire spread very quickly. So unless you *need* to use a door or window to escape, it's much better to shut them if you can.

WALLS LEAD TO EXITS

If you can't see – and you probably won't be able to – remember that if you follow a wall it will eventually lead to an exit: a window or a door.

So: follow the wall, using the backs of your hands – if you're going to get burned, better here than on the palm, which is more debilitating. If possible, move any furniture or obstacles out of the way as you go – again using the backs of your hands – rather than skirting round them.

DON'T USE LIFTS

Or any kind of chute or shaft. For two reasons:

1. Lifts can malfunction in a fire, leaving you between floors. Once that happens, you make the job of escape much harder.
2. Chutes and shafts act as chimneys. There will likely be a *lot* of fatal smoke.

Instead, use the stairs. Stairwells are bigger, so they won't chimney quite as badly. Make your way down them as quickly as possible.

AVOID BALCONIES AND ROOFS

They might seem tempting, but once you're on a balcony and the fire has reached the adjoining room, you've nowhere to go. Rooftop helicopter rescues aren't like they are in the movies. They are rare, difficult and dangerous.

KNOW THE RISKS OF JUMPING

You can jump relatively safely, though not without risk, from a first-storey window. From a second-storey window, we're talking broken legs and pelvises at the very least. Third-storey or higher is potentially fatal. Any jump higher than a first-storey must be a last resort.

Don't necessarily rely on trees to break your fall. They *can* have that effect, but you can easily impale yourself on upward-pointing branches.

You can mitigate the risks of jumping by:

- Protecting your head. Wrap anything around it that you can find. Towels, jumpers and sheets will all absorb accidental impact to a degree
- Softening your landing. Mattresses, cushions, pillows – any soft material will soften the impact. Throw them out first and land on them.
- Making a rope. Tie sheets, blankets or any good strong material together. Tie this to a fixed point such as a radiator or a heavy piece of furniture and use it to climb out of the window. Don't worry if it doesn't reach the floor: anything that reduces the height you have to jump increases your chance of surviving. If you can't make a rope, dangling from a window ledge before dropping will also reduce the impact when you hit the ground.
- Knowing how to land: with your knees bent and your weight going forward. As you hit the ground, roll out of the fall.

MOVING THROUGH FIRE

Sometimes, your only way of escaping is to move through the fire. If you have to do this, your best chance of survival is to soak yourself with water. So, if you can, get to a bathroom. Remove any heat-conducting items on your person – watches, jewellery, belts. Roll down your sleeves and button up your shirt to cover as much skin as possible. Now, drench your clothes and hair. Cover your head and face with whatever wet material you can get your hands on. This may give you the few extra seconds you need as you sprint through the flames.

THINK HARD BEFORE
TRYING TO RESCUE SOMEONE

It's commendable to want to try, but your chances of doing it are minuscule. You'll be lucky to get yourself out of a burning building, let alone drag out someone else who is unconscious.

This is one of those moments when you need to make some snap decisions about the priority of survival. Obviously if it's your children or loved ones, you'll do whatever it takes to save them. But understand that your chances of death are significant. Firemen will all tell you the same thing: entering a burning building to rescue someone, even with a respirator on, is a high-stakes game.

K.I.S.S.
(KEEP IT SIMPLE, STUPID!)

The smoke kills you long before the fire.

Close non-exit doors and windows –
more oxygen means hotter fire.

Stay low and follow walls – they lead to exits.

Only attempt a rescue if you are prepared to seriously
compromise your own survival.

HOW TO

SURVIVE

A CBRN

ATTACK

CBRN stands for 'chemical, biological, radiological and nuclear'. All first responders are trained in CBRN skills, for good reason: if you find yourself in the middle of a CBRN incident, you are in perhaps the most deadly modern survival situation. There will be deaths. But there are a few things you can do to make sure you and your loved ones are not among them.

Firstly, don't underestimate the likelihood of these events. Plenty of people go trekking in high-risk areas in countries such as Pakistan and Syria. There are rogue states in possession of nuclear and chemical weapons. And it's not a matter of *if* a terrorist cell manages a significant chemical or biological attack, it's a matter of *when*.

CHEMICAL AND BIOLOGICAL ATTACKS

The idea of a dirty bomb is a sobering one. In fact, bombs are not the most efficient way of delivering chemical and biological agents. They are more likely to be delivered, and dissipated, in the air above you, or by using improvised distribution systems on the ground. Many of these are extremely basic. Anthrax is one of the most deadly biological weapons, and it has been delivered to targets in envelopes through the post.

All this means that you may not know an attack has taken place until you or people around you have been affected. So you need to be aware of the symptoms.

CHEMICAL WEAPONS

These can be roughly divided into four different types: nerve agents, blister agents, blood agents and choking agents.

- **Nerve agents**, such as Tabun and Sarin, affect the central nervous system. Symptoms include a runny nose; a tight chest; constricted pupils; twitching, jerking limbs; spasms; suffocation.
- **Blister agents**, such as mustard gas, affect the skin. Symptoms include erythema (a reddening of the skin that looks a bit like sunburn), conjunctivitis, massive blistering of the skin.
- **Blood agents**, such as hydrogen cyanide, affect your blood's ability to move oxygen around the body. Symptoms include headaches, nausea, breathlessness, convulsions, respiratory weakness, cardiac arrest.
- **Choking agents**, such as chlorine gas, are the least dangerous but still seriously nasty. Symptoms include coughing, choking, nausea, vomiting.

BIOLOGICAL WEAPONS

These are weaponized versions of deadly bacteria and viruses. They tend not to work as quickly as chemical weapons, but they are at least as deadly. There are many different biological agents. Three of the scariest known to be weaponized are:

- **Anthrax.** Incubation time: one to five days. Symptoms
  include nausea, chest pains, respiratory problems,
  sepsis, bloody diarrhoea, skin lesions.
- **Botulinum.** Incubation time: one to five days. Symptoms
  include blurred vision, fatigue, respiratory paralysis.
- **Ricin.** Incubation time: eighteen to twenty-four hours.
  Symptoms include fever, respiratory failure, circulatory
  failure.

Add to this list such pathogens and toxins as smallpox,
pneumonic plague and viral equine encephalitis, and you've
probably got the picture: this is a very nasty class of weapon,
leading to horrible, painful ways to die.

WHAT TO DO IF YOU SUSPECT A CHEMICAL
OR BIOLOGICAL ATTACK

You have three survival priorities: evacuation, body protection and
immediate emergency medical care.

Biological and chemical agents generally can't be seen or
smelled, so you've no way of knowing if they're in the air around
you. Getting out of the vicinity is an immediate priority. Since
these agents are generally airborne, you should move at 90° to
the direction of the wind to remove yourself from the danger zone.

If you think you are at immediate risk of being exposed to a
chemical or biological agent, you need to cover your body. First
responders have special CBRN suits that cover their entire body
and head, with built-in respirators. You won't have one of these, so
you'll need to improvise. Tuck trousers into socks, get gloves on,
button up your shirt and put up your collar. If you have access to
swimming goggles, put them on. Wrap a damp cloth round your
face and cover yourself in whatever waterproof clothing, sheets or
blankets you have.

Anyone infected by any of these agents needs medical help
as quickly as possible. Some biological agents have vaccines, but
you're only going to get these from a medical professional. In any
case, you're going to need advanced medical care, and even that
might not be enough.

Avoid skin contact with other people. Certain categories of

biological agent are highly contagious and easily transmitted from person to person.

NUCLEAR AND RADIOLOGICAL ATTACKS

First off, you need to know what happens when a nuclear device explodes. The event can be divided into three components:

1. **The thermal pulse.** This is a short blast of blinding light, followed by a longer blast of intense heat – this can be millions of degrees at the epicentre and will vaporize anything within a certain radius, depending on the size of the device.
2. **The blast wave.** This is a high-pressure wave that forces air to move at extremely high velocities – easily fast enough to destroy buildings. It can rip away human skin burned by the thermal pulse. It's accompanied by an electromagnetic pulse that will disable all electronic equipment in the area.
3. **The radioactive fall-out.** The blast sends radioactive particles up into the air. These fall back to earth, like snow. There are three types of radiation: alpha, beta and gamma. Gamma's the worst. If you're exposed to high levels of it, you die in a few weeks.

If you're in the vicinity of the thermal pulse and blast wave, your chances of survival are very poor. You can increase them by:

- lying low on the ground, facing away from the blast zone
- keeping your mouth open, which stops pressure building up inside your lungs as the blast wave hits, and exploding them

If you're further from the epicentre, you have more choices. In this situation, your principal concern will be the radioactive fall-out. There will normally be a warning that a nuclear strike is to occur, so you have a call to make: run or seek cover.

WHEN AND WHERE TO RUN

The danger zone from a nuclear explosion is wide and varies according to the size of the bomb. The fallout from a 100 megaton bomb can travel hundreds of miles.

So if you're going to evacuate, you need to get a long way away. This will be difficult. There may not be time. The electromagnetic pulse will have taken out many modes of transport, and there will be general panic. Realistically, you'll need to seek cover at some point.

NUCLEAR SHELTERS

In an urban environment, you're looking for thick walls – preferably concrete – and no windows. If it's underground, that's much better. Try to get to a solid commercial building with a basement. If you have five storeys above you, you'll only be exposed to a two hundredth of the fall-out radiation. The ground floor of a wooden-framed house with windows, by contrast, will only protect you from half the radiation.

Whatever your shelter, however, you can improve your protection by surrounding yourself with dense, thick objects. Think furniture, filing cabinets, books: pile them up around you to make a shelter within a shelter.

Turn off all air-conditioning, and avoid water, which will likely be contaminated. Don't eat any stored food, because it will have absorbed some radiation.

In an outdoor environment, try to locate natural shelter: caves, ravines, outcrops. Otherwise, dig. Get in that hole as quickly as possible and erect some kind of roof – even if it's just an item of clothing. Now keep digging. Cover yourself with a thick layer of earth if possible.

RESIDUAL RADIATION

The worst of the radiation will have passed after forty-eight hours, but there is still a high risk of residual radiation. If possible, you should stay in your shelter until the emergency services arrive to relocate you to a safer area.

K.I.S.S.

(KEEP IT SIMPLE, STUPID!)

Know the common symptoms of exposure to chemical and biological agents.

Cover up to avoid exposure.

The best nuclear shelter is underground, encased in concrete, no windows.

Don't even think about leaving a nuclear shelter for at least forty-eight hours after the blast.

HOW TO

SURVIVE

A

KIDNAPPING

The threat of kidnap is a small but potential one in many parts of the world. Here are some of the most important countermeasures in kidnapping situations.

THE FIRST FEW SECONDS

Your best chance of thwarting an abduction attempt is in the first few seconds. You must do everything, in those moments, to fight off your assailant. That means:

- Fighting dirtily and aggressively. Go for the eyes, the neck, the genitals. Hit soft target areas with hard weapons. Hold nothing back.
- Screaming and shouting. Alert people to your predicament. Do not go quietly.
- Not being a soft target. You have a good chance of scaring off your abductor if your response is unexpectedly violent and crazed.

THE FIRST TWENTY-FOUR HOURS

If your initial resistance fails, however, your strategy must change. Know that the first twenty-four hours of a kidnapping attempt are the most important. You are likely to be moved several times in those twenty-four hours. The more often you're moved, the harder you are to find. Moreover, if you're to make an escape attempt, you're at your strongest earlier on in your ordeal. You'll have eaten and drunk water recently, you'll be in better physical shape and you'll be thinking more clearly. Your condition is only going to deteriorate with time.

WHAT KIND OF KIDNAP IS IT?

The most important information you can glean is the *reason* for your abduction. This will inform how much risk you should take in attempting to escape. If it's for ransom or leverage, remember you are much more valuable to your abductors alive than dead. In these instances, you should only attempt to escape if you have a high probability of success. Remember that most people kidnapped under these circumstances survive the ordeal.

If you've been abducted by a sexual predator or for retaliatory political/military reasons, however, your life is almost certainly at risk. Take your chances early. Be alert for the signs that your abductor is planning to kill you. These include:

- brutal treatment
- a frightened or desperate demeanour
- the release of other hostages but not you
- your abductors revealing their identities having previously kept them secret by wearing masks

In these instances, you *must* attempt to escape even if your chances are not good.

PRE-ESCAPE INTEL AND DEMEANOUR

Escaping is hard. Your priority should be to feign compliance, because you don't want your abductors to think that you need extra restraint. Avoid eye contact, keep your head down and appear submissive. If your abductors don't think you're going to fight back, holes are more likely to creep into their security.

Try to keep your abductors calm. There is no benefit to you in making them angry, nervous or violent.

Try to build up a relationship with your kidnappers. Don't insult or disagree with them. Be empathetic. It will be harder for them to harm you if you make yourself seem more 'human' and establish some kind of bond with them. Ask for small favours: a glass of water, something to eat, a newspaper. Give your abductors the opportunity to humanize you and to be humanized in return.

While you're establishing this bond, however, you must continue to be constantly observant. In the early minutes and hours, try to make a mental note of the route the kidnappers take – even if you're blindfolded in the boot of a car, you can keep a tally of left and right turns you've taken, and count the time taken between them. When you reach the place where you're being held, be hyper-alert.

- Where are the potential exits?
- Are there any obstacles?
- How many abductors do you have?
- What is their routine?
- Are they armed?
- Do they seem fit?
- Is their mental state volatile?

The more intel you have, the higher your chances of success.

ESCAPING RESTRAINTS

The two most common means of restraining someone are zip (or cable) ties, and duct tape. You can escape from both of them.

ZIP TIES

If you look carefully at the locking mechanism on a zip tie, you'll see that it includes a little ratchet that locks against the teeth of the tie. If you can lift this, the zip tie will unlock. The best tool for doing this is a hairpin, but it can also be done with a fingernail, a pin, a pencil or the corner of a credit card. Obviously this is easier to do if you have a helper, but it's possible, if fiddly, by yourself.

If you can't find a suitable tool, you can break a zip tie round your wrists. To do this, tighten it as much as you can by gripping the end of the tie between your teeth and pulling up (this might seem counter-intuitive, but it will make it easier to break). Raise your hands above your head then bring your wrists sharply down to your stomach while yanking your elbows outward. The zip ties should break at the locking mechanism.

You can also use your shoelaces to get free. Untie one of them from your shoe and make two loops, one in each end. Thread the shoelace either side of your zip tie, then place the loops over your feet. With a sawing motion, move your feet back and forth. This will generate friction heat and melt the cuffs.

DUCT TAPE

This is cheap, easy to obtain and very strong, which is why it's so often used in kidnapping scenarios. However, it can be defeated because its cloth backing makes it easy to tear. Strong sudden movements are key, because they will make the tape tear rather than crumple.

If your ankles are bound, stand up and align your feet in a 'V' position. Squat down as quickly and heavily as you can, aiming your bum to your heels and pointing your knees out. The tape should rip.

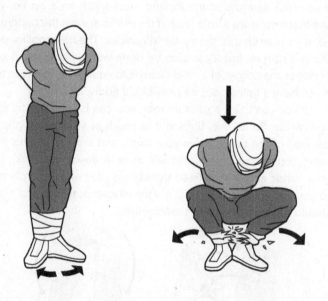

If your wrists are bound, point your arms forward and straight, then yank your elbows back past your abdomen. Again, the tape should rip apart.

RESCUE ATTEMPTS

If you're lucky, there will be a rescue attempt. This is statistically the most dangerous moment in your ordeal, after the initial abduction. You might be used as a human shield. Your rescue team will almost certainly be using firearms and explosives, and they will not necessarily know who is a hostage and who is an abductor. They will be on edge, and liable to fire at will. You need to:

- get down on the ground
- cover your head
- if you can, find protective cover
- do exactly what you're told and be prepared for some rough treatment – even to be tied up while they establish who's who

K.I.S.S.
(KEEP IT SIMPLE, STUPID!)

In the first few seconds of a kidnap attempt, be
hyper-aggressive and violent to thwart it.

In the first twenty-four hours, feign compliance,
but gather as much intel as possible.

Identify the type of abduction before judging
how much risk to take in escaping.

Learn how to escape restraints.

HOW TO

SURVIVE

A BEAR

ATTACK

When I hiked through Alaska with President Obama (not a sentence I thought I would ever write!) I found – and we ate – a raw salmon that had been half-eaten by a bear. It was proof that there were bears in the vicinity. On that occasion we had the benefit of sixty armed secret service personnel protecting us. Most of us are not going to have that advantage, so here's the low-down on how to conduct yourself in bear country, and what to do if things turn bear-shaped.

Bear attacks are pretty rare, but when they do happen, they can be catastrophic. A bear can easily kill a human, and it's a nasty way to go. In the wake of one bear attack in New Jersey, human remains were later found in the bear's stomach and oesophagus, and human blood and tissue were found underneath its claws. It doesn't take much imagination to work out how that encounter went. Your chances against a 1,500lb grizzly aren't great. They're bigger, stronger and faster than you are. So if you find yourself coming up against one of these potentially deadly animals, you need to play to your strengths. The one advantage you have over a black, brown or polar bear is your brain. Think smart and keep calm, and you've got a chance of getting out alive.

PREVENTING A BEAR ATTACK

One of the first rules of survival is that prevention is better than cure. If you're hiking in bear country, you need to take a few precautions to avoid getting into a potentially life-threatening situation.

- If possible, travel in groups. More humans, less risk.
- Make some noise – if bears, or other wild animals, hear you coming, they're likely to avoid you.
- If you make camp, keep things clean.

This last point is very important. If you leave food or rubbish lying around, you'll attract scavenging bears. It's a good idea to have your sleeping area, your washing area and your food-storage area a good 50 metres apart from each other, because then you're separated from the areas most likely to attract bears. Any unopened food should be suspended from a branch at least 5 metres off the ground. (Menstruating women should remember that the same goes for used tampons – grizzlies can't resist them . . .) All cooking, food-hanging and washing areas should not only be away from camp, but downwind – so you'll need to know the direction of the prevailing wind (see page 73).

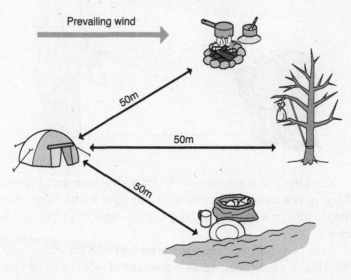

Prevailing wind

50m

50m

50m

BEAR SPRAY AND OTHER DETERRENTS

Bear spray – or 'pepper spray' – is by far the most effective deterrent in the case of a bear attack. It contains higher quantities of capsaicinoids than the pepper spray that is used against humans, and is *much* more effective against aggressive bears than a gun. A bullet is small and a charging bear easy to miss, whereas bear spray has a 20-foot spread at a distance of 25 metres. You can often hire tins of the stuff if you're hiking in bear country.

Don't just stick it in the depths of your pack, though. Have it close to hand at all times and ready to use. On your belt loop or in a jacket pocket is better than in a pack that you might not be close to at camp, for example. Like a soldier always keeping his weapon at 'arm's length', adopt the same discipline when in bear country.

If you do find yourself in the position of having to use bear spray on an attacking bear, you need to hold your nerve and only spray when the bear is within range. Have a practice squirt before you leave so you know what to expect. (Test-fire only when aiming downwind. I remember test-firing with President Obama and the wind suddenly changed and it almost all went very wrong!)

Consider a shotgun over a rifle, or even a flare gun, permits allowing. The spread of the shot from a shotgun will be much more likely to hit a moving, approaching target – especially when you're scared.

Take portable perimeter electric fencing to erect around your camp site. If a bear approaches in the dead of night this might not stop it, but it will alert you and at best will give the bear a shock that might deter it.

Maintain a fire and a fire-watch if you are in particularly intense bear country. This means taking shifts through the night to keep the fire roaring. Make sure you have collected a good supply of wood before it is dark. Nature often knows best, and a big fire goes a long way towards deterring bears and other animals. When it comes to bear attacks, prevention is *always* better than cure!

WHAT NOT TO DO WHEN A BEAR ATTACKS

- *Don't* try to run away. You have zero chance of outrunning a bear.
- *Don't* try to climb a tree. Bears, especially black bears, are better at it than you.
- *Don't* necessarily play dead (unless you're in the situation outlined below). Brown bears especially, when acting aggressively, are often after carrion. If you play dead, there's a good chance they'll see you as just that.

IDENTIFY THE TYPE OF ATTACK

So, you haven't been able to avoid a bear. You don't have any bear spray. The situation is clearly about to turn ugly. You know what you shouldn't do, but what *do* you do? How do you survive?

The key to maximizing your chances during a bear attack is to identify the bear's behaviour: defensive or aggressive?

DEFENSIVE BEAR ATTACKS

A female bear guarding her young probably isn't hunting you – she just wants you to get the hell out of there. If you come across a bear suddenly and unexpectedly, it's likely only to want to defend itself. The rule of thumb is that any attack that follows a sudden, surprise encounter is probably a defensive attack.

There are also certain warning signs to look out for. It is probably a defensive attack if the bear:

- sticks out its lips
- makes huffing or woofing sounds
- slaps the ground with its paws
- clacks its teeth (a sign of nervousness)
- rears up on to its hind legs (it's trying to see and smell you better, and find out more about you)
- has its ears laid back while charging

In this situation, you should back away, talking calmly and waving your arms so the bear identifies you as a human and not as some other threat.

If a defensive bear continues to advance on you, and certainly if it makes contact with you, playing dead is probably your best bet. The safest way to position yourself is in the foetal position, with your arms covering the back of your neck for protection. The bear might swipe at you, or even bite you, but there's a good chance it will then leave you alone when it realizes that you are not a threat. (If you're wearing a backpack, make sure you keep it on as it will give you an extra layer of protection.)

AGGRESSIVE BEAR ATTACKS

If the bear has been stalking and circling you and sees you as food, or is defending a kill site, it is most likely acting aggressively. It is probably an aggressive attack if the bear:

- makes very little noise
- displays none of the warning signals associated with a defensive attack
- has its ears pointing up and forward
- is maintaining eye contact while advancing on you
- is attacking you in your tent

In this situation playing dead is no good. Your only option is to fight back. In more open spaces you could, if possible, try to create some sort of diversion – maybe by throwing your backpack in its path so it stops to investigate. In the absence of bear spray, however, use anything you can as a weapon – fallen branches, rocks, whatever you can find. Make a *lot* of noise. Attack as if your life depends on it, because it does: once he's got his jaws and claws on your throat, it's a battle you're never going to win. Your aim is to encourage the bear to think that you're not an easy proposition, because it will keep attacking until it is too scared or injured to stop.

If you can't determine between an aggressive or a defensive bear attack, the rule of thumb is that if you stumble across a bear, you should back away, whereas if a bear encroaches on to your territory, you should stand tall, make a lot of noise and try to deter it from approaching further.

But if you are in bear country, be prepared: bear spray, a shotgun or flare gun (again, permits allowing) and portable electric perimeter fencing are all good deterrents. Trust me, you need all the help you can get, as this is one fight you want to do all you can to avoid.

K.I.S.S.
(KEEP IT SIMPLE, STUPID!)

In bear country, always carry, and be ready to use, bear spray and other deterrents.

Don't try to outrun a bear, or escape by climbing a tree.

Defensive bear attack? Back away calmly. If the bear makes contact, play dead.

Aggressive bear attack? Fight like your life depends on it, because it does. Improvise a weapon from rocks or sticks.

HOW TO

SURVIVE

A CROC

ATTACK

Crocodilians – that's crocodiles, alligators and caimans – kill hundreds, perhaps even thousands, of people a year. Mostly these attacks occur in Africa and Asia, where they often go unreported, but you'll also find these reptiles in Australia, South America and in the southern United States.

The most dangerous and aggressive crocodilians are saltwater crocodiles – 'salties' – and Nile crocodiles. But all of them are a potential threat. Crocs and their kin will eat pretty much anything. They also defend their territory very aggressively. If you're in croc country, you need to know how to avoid them, escape them and deal with them should they attack.

DANGER ZONES

Crocodilians can live in fresh or salt water – salties can be found on beaches and even in open sea. They will normally avoid areas of white water, but not always. Generally, though, they prefer slow-moving, muddy water with lots of vegetation. Think swamps and marshes, but also rivers and ponds. Even small ponds.

I remember once in the Northern Territory of Australia coming across a small pond of still water covered in a thin layer of muddy debris. We were a long way from any other water and the team thought it should be fine to wash in. I was less sure. I threw a rock in and almost before it hit the water a huge croc exploded out of the depths, snapped at the stone and disappeared in a burst of thrashing and waves. It was a sober reminder that crocs only need to get lucky once and that they can survive for a long time between mealtimes! They are the ultimate stealth predators.

So: if in doubt, stop, think and assess. Even throwing a few rocks into murky shallows before filling up water bottles will help ensure there isn't a monster waiting close by.

Generally, croc-infested areas are well marked by warning signs. But not always. Remember that crocs are very well adapted to merge into their environment. Just because you can't see them, it doesn't mean they're not there.

And don't always trust 'locals' who say an area of river or billabong or coast is croc-free. They don't always know for sure. Many a person has lost their life on duff intel from locals. If in doubt, stay out!

DANGER TIMES

You need to be on high alert at all times in croc country. But crocs are *most* active in the evening and at night. (A torch will help you identify them, because it will reflect off their eyes.)

You need to be extra careful during their breeding seasons. These vary according to the type of croc and their location. Arm yourself with local knowledge before you set foot in croc country.

HOW CROCS KILL YOU

You might have heard of the 'death roll'. This is when the croc, gator or caiman grabs its prey then rolls it repeatedly and violently in the water. Depending on the size of the beast, this can rip off chunks of flesh or entire limbs. Not that it matters, because the death roll will drown you anyway. You'll then be dragged down below the water line and wedged under a log till your flesh starts to rot and you can be easily torn apart and eaten in small pieces.

AVOIDING A CROC ATTACK

Avoiding a croc attack is a matter of knowing not only how but also when they attack:

- Although crocs will attack an adult human, they prefer smaller prey. Children and dogs should never be allowed near the water in croc country.
- Crocs have brilliant camouflage. They can hide underwater for long periods waiting for their prey to come to them (their heart rate goes down to two to three beats per minute). So never linger by the shoreline in croc country.
- They're clever. If you go to the water to drink, they'll take note. Go to the same place for a second time, and they'll strike. So don't take water from the same place twice.
- Crocs can launch themselves vertically from the water. In croc country, avoid low bridges, don't lean over the sides of boats and don't hang from tree branches over the water.
- *Never* approach a croc nest. Crocodilian mothers will do whatever it takes to protect their eggs and their young.
- If you find yourself in croc- or gator-infested water, as I once did, swim underwater. It's about the scariest thing you'll ever do, but it stops the crocs mistaking your bobbing head for easy prey like a bird or turtle.

WHAT TO DO IF YOU SEE A CROC ON LAND

If you spot a croc, gator or caiman on land, you have a pretty good chance of survival provided you don't get too close. They tend not to chase their prey, preferring the element of surprise – it's when you *can't* see them attacking that you need to worry. So:

- Back away slowly. Don't approach it. Crocs rarely attack when they're on land.
- If they are heading towards you, and seem to be acting aggressively, run. They're much slower on land than they are in water (where they're *fast*). Their top speed is about 10 miles per hour, and they tire quickly, so you should be able to outrun them.

The old survival tip of running in a zigzag when you're escaping a croc isn't much use. Humans are generally faster on land, and running in a straight line will put more distance more quickly between you and the croc.

WHAT TO DO WHEN A CROC ATTACKS

So what if the worst happens, and a croc gets its jaws round you? This is a situation that most people don't survive. But there are ways of improving your chances.

If a croc bites you on land and then releases, it's acting defensively. Run away as quickly as possible. (Remember: no zigzags.)

It's much more likely, however, to be an aggressive attack. If you're bitten on land, the croc will probably try to drag you to the water. You need to attack with every last bit of strength you have. Your only chance is to stop the croc thinking you're easy prey. So:

- Go for the most vulnerable parts. You'll never pierce the skin, especially of a large crocodilian. Their weakest spot is the eyes. Gouge, pierce: do whatever you have to do.

- If you can't get to the eyes, strike the head as viciously as you can. It's still thick-skinned, but most animals instinctively want to protect that part of the body.

- If you have no other option, there is a palatal valve at the back of the crocodile's mouth. This stops water going down its throat when it is submerged. If you can punch that, water will flood into its lungs. It'll have to release you, or it will drown. (Note, however, that at this stage you are in big trouble.)

- If you're caught in a death roll, try to roll with the croc to avoid it tearing your limbs from their sockets. Do whatever you can to avoid being dragged into the water. Even if it has torn a chunk of flesh from you, or even a limb, consider yourself lucky and get away – it might be enough to satisfy the croc. Avoid being in the water *at all costs*.

GET TO THE MEDICS

If you've survived a croc bite, you'll be in a bad way. Think blood loss and badly ripped flesh. There is also a very strong chance of infection. Crocodilian mouths are teeming with bacteria. These can kill you as surely as the bite itself, especially in warm, humid areas.

K.I.S.S.

(KEEP IT SIMPLE, STUPID!)

Stay clear of the water in croc country: crocs are highly adapted predators, well camouflaged and cunning.

Don't be tricked into thinking a body of water is croc-free.

If a croc is coming for you on land, run – you're probably faster. And don't zigzag.

If it bites, attack its eyes and head violently. Do whatever you can to avoid being dragged into the water.

Last-ditch attempt: punch the palatal valve at the back of its throat.

HOW TO

SURVIVE

A DOG

ATTACK

Don't underestimate dogs, whether wild, domesticated or trained. In Africa, I would describe wild dogs as the most efficient hunters in the whole food chain. Special forces soldiers use dogs as weapons. These awesome animals are trained to jump from planes, are often fitted with special body armour, and even have metal coating applied to their teeth to make their bite more effective. Dogs like that are smart, strong and potentially deadly in a fight.

Aggressive canines, even ones that aren't trained as attack dogs, can and do kill. Knowing how to deal with them can mean the difference between life and death.

IS THE DOG DANGEROUS?

Some dogs are statistically more dangerous than others. Pit bulls, Rottweilers and German shepherds cause the most injuries to humans. But any dog can be dangerous. If it has been badly treated, regularly chained up or confined, or feels that it is under threat, a dog is more likely to be aggressive. You should never tease or taunt a dog, especially one you don't know.

WARNING SIGNS

Pay very good attention to the dog's body language. Signs of aggression include:

- growling or baring its teeth
- barking
- head in line with the body
- raised hair
- narrowing of eyes

If a dog *is* showing signs of aggression:

- Don't look it in the eye – it will take that as a challenge.
- Don't smile – it could interpret that as you baring your teeth.
- Stand sideways to it so you present less of a threat.
- Gradually back away.
- Make no sudden movements as you get yourself out of the dog's line of sight.

WHEN TO RUN

If backing away doesn't work, and you sense that conflict is imminent, you need to make some fast, difficult decisions.

In general, you *shouldn't* try to run away from an aggressive dog as this triggers its chase instinct. But there are exceptions to this rule. If a number of dogs are attacking you, your chances of winning that fight are small. In this case, your best chance of getting out alive is to remember that dogs can't climb. Run, if necessary, and get up high. Climb a fence. Jump on a car.

If that's not possible, remember that dogs *can* be intimidated. Raise your arms above your head to make yourself appear as tall as possible, and roar loudly. If there are stones nearby, throw them hard at the dog to make it think twice about attacking.

DEALING WITH AN ATTACK

If the dog keeps coming, its primary weapon will be its bite. Do what you can to get something between you and that bite – a stick, a jacket or a rucksack maybe.

If the bite is unavoidable, you need to think carefully about which part of your body you would prefer the dog to attack. It's a Hobson's choice as any kind of bite is going to hurt, but you want to stop the dog from attacking your most vulnerable parts: your neck, your face, your chest and your groin. This might mean offering it another part of your body. If possible, you should wrap whatever clothing you can around your forearm and offer that to the dog. Your arm has a higher proportion of bone than your leg, and will bleed less. This manoeuvre will leave you with three limbs with which to fight back.

Remember, though, that when dogs bite, they dig their teeth in firmly and don't let go. If you try to rip your arm from that bite, all you'll do is tear your flesh. You'll end up with a vicious, debilitating open wound.

Offer your protected arm and, when the dog leaps, it is better to turn with the jump rather than fight the momentum head on. Once a dog has you in a bite, keep it on there and administer your counter-attack until it releases or scampers away wounded. You want to avoid a ground fight, because this exposes your vulnerable parts, so stay on your feet with a solid, wide, low stance and manoeuvre to maintain your balance. The aim now is to neutralize the animal as quickly as possible.

With your free arm, go for the dog's eyes: thumb and middle

finger into the eye sockets and squeeze hard together. The goal is to blind, maim or kill. You can also strike at the back of its head, just at the base of its skull, with a small rock or whatever you can find around you. If you're in the wild and you have a survival knife with you, the dog's most vulnerable parts are under its front leg or just above its shoulder. Other tips include covering the dog's head with a coat, which often subdues it, and lifting the dog's hind legs up in the air, which stops it from manoeuvring effectively, then twisting them violently until either a leg or the pelvis snaps or breaks.

One other way to neutralize an attack is to use your body weight and fall on the dog to crush it – a dog's ribs break easily.

MULTIPLE DOGS

If you are being attacked by more than one dog, and you haven't managed to run from them, techniques such as falling on one of them and using your body weight won't work, because you'll be vulnerable to attack from the others. In this circumstance you need to use their pack mentality against them. Try to hit the nose and eyes of each dog as it approaches. Use controlled stamp kicks, aimed at their heads. If possible, grab their limbs and pull them in opposite directions to break or maim. If the pack senses that you are a tough proposition, there is a good chance that it will retreat.

TREATING DOG BITES

A bad dog bite will probably need stitches. In the short term, you need to stem the bleeding as quickly as possible, but in the long term you're going to need proper medical attention.

Torn flesh isn't your only problem. One in five dog bites, no matter how severe, results in infection. That can kill just as surely as the dog attack itself and is another reason why, if you've been bitten, you need to get yourself checked out at a hospital.

K.I.S.S.
(KEEP IT SIMPLE, STUPID!)

Never assume that certain dogs can't be aggressive.

Running from a dog triggers its chase instinct –
only do it if you're being chased by several.

If a bite is inevitable, wrap your arm and offer that.

Fight to stay on your feet, then attack the dog's vulnerable
parts, such as eyes and nose, limbs or groin.

HOW TO

SURVIVE

A SHARK

ATTACK

Humans are a far greater danger to sharks than sharks are to humans. We kill more than a hundred million a year. The way we treat them is a tragedy – they're an important part of our ecosystem, and we should be trying to protect them. Most sharks are not aggressive to humans, so your chances of being attacked by one when you're swimming from the beach are pretty low.

For the survivor at sea, however, it's a different matter. When the USS *Indianapolis* was hit by a Japanese torpedo in the Pacific in 1945, some nine hundred men were stranded in the water. Over the next four days, two thirds of them were ripped to shreds by ocean sharks. Only 317 survived.

The sharks responsible for the most human casualties are the great white, the mako, the tiger, the bull, the grey nurse and the hammerhead. They can be ferocious, deadly creatures. A shark's jaw exerts 20,000kg of pressure per square inch (humans: 150kg per square inch). They can easily rip through flesh and bone, and there are many recorded instances of sharks tearing a human body in two at the first strike.

Sharks are highly evolved natural predators, and they're stronger than you. If you find yourself at risk of a shark attack, your only hope is to use your superior brain.

PREVENTING A SHARK ATTACK

As many people know, sharks can sense blood in the water from three miles away. However, they can also sense waste products. If you're stranded at sea and have to urinate, do it in short bursts so the urine can disperse. If you need to defecate, grab the faeces as it comes out and throw it as far away from you as possible.

If there's more than one of you in the water, huddle together. Sharks are less likely to attack a group.

If there are sharks in the vicinity, give them plenty of space and avoid provoking them. Even a non-aggressive shark may attack if it feels cornered.

READ THE SIGNS

If sharks are nearby but not approaching, and are swimming slowly and smoothly, they are probably just curious and checking you out.

Warning signs of aggression include:

- circling
- pointing their pectoral fins down
- pointing their heads upward
- arching their backs
- zigzagging
- charging

Side

Front

Top

Aggressive
behaviour

Non-aggressive
behaviour

WHAT NOT TO DO

Do not scream, flail or panic. Easier said than done, but terror is music to the ears of even a less dangerous shark. The predator instinct is hard-wired in them. If they detect weakness they will see you as prey and will be *far* more likely to attack.

SHOW CONFIDENCE

If you find yourself in water where there are dangerous sharks around, your best chance of survival is to disrupt their instinctive behaviour. If there is a boat nearby, swim towards it with as much strength and confidence as you can muster. But don't rush. Move gracefully and try not to splash – the shark might interpret this as panic.

If there is no boat, and you are being circled by an aggressive shark, your best chance of survival is to match its aggression. As counter-intuitive as it might seem, it has been shown that if you swim confidently towards the shark, shouting at it underwater, it will likely turn away at the last minute. You are showing the shark that you are not its natural prey. It will make the predator uncertain. When they are uncertain, they tend not to bite.

A friend of mine is a naturalist who studies great whites. He confirms that this is the best strategy. Whenever he is swimming with sharks, and one of them starts to get a bit frisky, he swims towards it. It confuses them, and stops them from becoming aggressive. On one occasion he didn't notice that a shark was attacking him from behind and below. He turned at the last minute, and his reflex reaction was one of panic. He knew, though, in that moment, that if he continued that reaction, and the shark sensed his distress, he'd be in real trouble. With huge self-control and confidence, he had the presence of mind to turn and swim directly at the beast. It swam away at the last second. As so often happens in survival situations, he needed to display total belief and total commitment in the big moment to make it through.

WHAT TO DO IF YOU'RE BITTEN

First, remember that great whites in particular (and they account for a third to a half of all shark attacks) are naturally curious. Humans are not their natural prey, and if they bite, it is often a 'sample bite' and there is a good chance they will release you. You'll be in a desperate situation, you'll be a mess, but there's a chance you'll survive.

If a shark has you in its grip, you need to go for its most vulnerable points. These are the gills and the eyes. You need to punch the shark as hard and aggressively as you can in these areas. If you have a knife, that's where to stab it. You'll hear stories about people trying to strike them on the nose, but this is not a particularly weak spot, and it is dangerously close to the teeth.

Some sharks, including tigers and hammerheads, will enter a state of paralysis (it's called tonic immobility) if you turn them upside down. Not easily done, but worth a try if it has a hold of you.

SHARK REPELLENT SPRAYS

It has been shown that sharks tend to avoid areas where there are dead sharks. Shark repellent sprays are made from bits of putrefied shark, and claim to cause a violent bolting reaction if you spray them in a shark's direction. If you're going to be in shark-infested waters, by all means take some of this stuff, but remember: you only get one go, so wait until you definitely know you're under attack, and the shark is close.

K.I.S.S.
(KEEP IT SIMPLE, STUPID!)

Sharks are attracted to blood, faeces and urine.

Most sharks are not aggressive, but if they aren't swimming smoothly they could be about to attack.

Your best defence is confidence. If you swim towards an aggressive shark, there's a good chance it won't continue its attack.

If it has you in its grip, punch or stab its most vulnerable points: the eyes and the gills.

HOW TO

SURVIVE

A SNAKE

ATTACK

Although I've spent a lot of time in snake-infested areas, I've only been bitten once. That was in the Borneo jungle. I had made a small tree camp, night was falling and I was hungry. I saw something slither along a branch. Sensing supper, I grabbed it by its tail. It wrapped itself firmly round the branch so I pulled harder. When it eventually pinged free, it flew back and bit my hand.

I was lucky. That bite was non-venomous. But even a non-venomous bite can be dangerous. Soldiers in the jungle are always reminded that 'snakes don't brush their teeth'. Their mouths are full of bacteria, so a snake bite can easily cause a dangerous infection – especially in warm, humid areas. So don't do what I did and antagonize a snake unless there's a genuine survival reason for it.

AVOIDING SNAKES

Generally speaking, snakes will avoid you – they are shy creatures and want as little contact with humans as possible. There are exceptions, though. I've come across very aggressive water moccasins in the swamps of Louisiana, for example, that have actually swum straight towards me intent on biting. I had to hit one very hard with a stick!

If you're in a snake-infested area, you should take the following precautions:

- Walk carefully, but with a slightly heavier footfall – snakes can sense vibrations from a long way off (they actually hear with their jaws).
- If you are walking through thick undergrowth, use a sturdy stick to clear the ground in front of you.
- Wear sturdy jungle boots and snake gaiters.
- Step on top of logs first, rather than go straight over them. Snakes shelter under the curves of a log and if you simply step over, there's a high risk you'll be bitten.

HOW SNAKES BITE

Venomous snakes often deliver what is called a 'dry bite', which means that they don't inject any venom. This usually happens when they are in poor condition, or have recently envenomated, but it also happens when they are issuing a warning.

Younger snakes, especially baby rattlesnakes, can be more dangerous than adult snakes because they can't control the amount of venom they produce. They generally envenomate fully.

IDENTIFYING VENOMOUS SNAKES

There are lots of old wives' tales about how to identify venomous snakes. None of them are reliable. If you know you're going to be in a snake-infested area, take time to learn what snakes you are likely to encounter, and how to identify them. Even then you should treat *any* snake bite as potentially venomous.

If you are bitten, these are the features you need to try to remember in order to identify the snake for a medic:

- length
- thickness
- colour and patterning
- shape of the head (triangular or not?)
- shape of pupils (round or slitted?)

If you or a buddy can take a picture, do so. But don't try to kill the snake yourself – it wastes time, and by exerting yourself you move the venom round your body more quickly.

The most common symptoms of a venomous bite are:

- swelling
- difficulty breathing
- nausea and vomiting
- blurred vision
- sweating
- numbness

TREATING A SNAKE BITE

Ignore all that stuff you see in the movies about sucking the venom out of a snake bite. It's a very bad idea. You'll just open up more capillaries around the wound, and there's a good chance of getting venom in your mouth. Don't try to cut the venom out, either – you'll end up getting it into your bloodstream faster.

If you're bitten by a snake, these are the steps you need to take:

1. Stay calm. Easier said than done, but you need to keep your heart rate low to stop the venom spreading. The more you panic, the higher your heart rate and the faster the venom moves around your body.

2. Keep the bite below the level of your heart.
3. Wash the wound.
4. Place a constricting bandage – but not a tourniquet – above the bite, to prevent the venom from spreading.
5. Get to the nearest hospital swiftly but calmly.

The last step is the most important. Whenever you are heading into an area where you know there are venomous snakes, you need to have a solid casualty evacuation (casevac) plan. If you have a bad haemotoxic or neurotoxic bite, you're going to need a hit of the right anti-venom and the sooner the better. However . . .

ANTI-VENOM
Anti-venom is not the magic pill everyone thinks it is. It may well save your life and the faster you can get this treatment the better. Every minute really does matter. But if it's a bad bite from a very venomous snake, you're still at risk of losing a limb, or worse. They can also produce severe side effects.

Even when you do everything right, and you survive, snake bites can be very ugly, as a friend of mine found out in the jungle of Costa Rica. A fer-de-lance bit him through the eyelet of his jungle boots. (That was an unlucky one!) He got to hospital within two hours, but even by then the venom had caused extreme muscle necrosis – it was eating my friend's muscle away. The surgeons eventually had to cut off the top of his foot, remove his quad and staple that in its place. (Ugly sight, but it saved his life and his foot.)

HOW TO KILL A SNAKE

The best way to kill a snake, if you really have to do it, is with a heavy club or stick. You should aim for its back, just below the head. If you can, pin the snake down first with a forked stick, then administer the blow. Wounded snakes are extremely aggressive, so try to make your first blow count, then sever the head. The head can still envenomate after death, so you should burn or bury it carefully.

K.I.S.S.
(KEEP IT SIMPLE, STUPID!)

In snake-infested areas, wear proper boots and snake gaiters.

Use a heavy footfall and move the undergrowth with a solid stick.

If you're bitten, never try to suck out the venom or cut the wound.

Keep calm, keep the bite below the level of the heart,
and get to a hospital quickly.

MEDICAL EMERGENCIES

HOW TO
AVOID
BLISTERS

There's an old army saying: a soldier is no better than his feet. It's true of anyone in a survival situation. Our feet are very prone to small but debilitating problems. They're also our principal form of transport. It doesn't matter how fit you are, or how well prepared for an extreme environment, often our feet are our limiting factor. If they can't do their job, little else matters.

So while a blister might sound like a tiny thing, if you're relying on your feet to keep you alive, it could easily be the death of you. You need to know how to *avoid* them and *treat* them.

PREPARATION

Preparation is the most important aspect of foot care. If you can get your feet rugged and hard, you'll be far less prone to blisters. There's only one way to do this. Get out there. Walk. Train. The more you do it, the harder your skin will become in all the places where it needs to be tough.

New walking boots are blister machines. I often see people with a shiny brand-new pair of walking boots at the start of a three-month trek, and I know they're going to have trouble. I've experienced that myself during simulated basic training with the French Foreign Legion. We were given new leather boots on our first day and ordered to walk for miles across the desert. By the end of the day our feet were bloodied and our group of tough guys were hobbling like old ladies.

So: if you're heading out into the wilderness, it's crucial that you wear your boots in first. They need to soften up, mould to your feet and get wet and dry several times. A good trick is to put them on, lace them up and stand in a bowl of hot water for a couple of minutes. Then walk around in them for a while and let them dry.

Wear good-quality socks and change them often. Sweaty, damp socks increase your chance of a blister. Looking after your feet at the end of the day is essential. Take wet socks and boots off, clean your feet, if possible apply talc (Marines call it 'foo foo powder'), and put on dry socks.

HOT SPOTS

Even when our shoes fit perfectly, and our soles are tough and hardened, we can sometimes be prone to blisters. So we need to be hyper-aware of the 'hot spot' that precedes them. You've probably felt it before: a localized burning sensation on your foot while you're walking.

Whatever you do, don't ignore that hot spot. If you don't deal with it immediately, you're going to get a blister.

On page 22, you'll see that one of my essential pieces of kit is a roll of zinc oxide tape. This is a super-sticky fabric tape. Climbers use it to wrap around their fingers for protection and grip. We're going to use it for blister prevention. Take your shoe off, find that hot spot – it'll probably be red – and slap on a strip of zinc oxide. The tape works round the back of your heel, under the arch of your foot and round your toes. Keep it there as long as you need to. I once had the same strip of zinc oxide stuck like superglue on my feet for a month. It's a bit skanky when it comes off – but no blister.

WHAT IF YOU GET A BLISTER?

In an ideal world you would never pop a blister, because the liquid over the wound is your body's way of keeping it sterile. This isn't always possible, especially in a survival situation when you need to keep moving. They're probably going to burst anyway, so it can be a good idea to burst them with a needle, if you have one. Sterilize it first in boiling water or over a hot flame. Massage the liquid out of the blister, but keep the dead skin in place as a layer of protection.

Alternatively, if you have them with you, use the needle and thread trick. Thread a sterilized needle. Pierce it in one side of the blister and out the other side. Leave half a centimetre of thread hanging from either side of the blister. This will wick the liquid away and allow the blister to drain.

A blister is a wound like any other and can easily become infected, especially in the hot, humid environment around your sweaty feet. So you need to clean a blister regularly, especially if it has burst of its own accord. Clean water is fine. Antibiotic lotion or rubbing alcohol is better. Be on the look-out for smelly pus, swelling, heat and spreading redness. These are all signs of infection. Once your blister is clean, you can cover it with a plaster or tape. Do everything you can to keep your socks dry.

In extreme cases you can squirt some superglue into the burst blister. It'll hurt like hell, but will form a hard protective covering for the wound.

K.I.S.S.
(KEEP IT SIMPLE, STUPID!)

Tough feet get fewer blisters. Harden them up.

Don't go out in the field with new boots.

Hot spot? Get zinc oxide tape on it quickly.

Keep a burst blister scrupulously clean to avoid infection.

HOW TO

TRIAGE

IN AN

EMERGENCY

In an extreme survival situation there is a high likelihood of severe, life-threatening injury. We're not talking cuts and grazes here. The information I'm giving you in this book is for emergency situations when professional medical care is unlikely. You might have to make some fast, difficult decisions.

Sometimes the first aid responses can be as gruesome as the injuries themselves. But there's no time for squeamishness when you have to do whatever it takes to keep someone alive.

In an extreme medical situation, when you might be presented with horrific quantities of fast-flowing blood, ear-splitting screams, burns that peel away multiple layers of skin or breaks and dislocations that result in bones at horrible angles protruding through the skin, it can be easy to panic.

But in these scenarios, panic is a killer.

That's why it's essential to have systems in place for prioritizing the needs of the wounded, and for treating the right injuries, in the right way, in the right order.

That's where triage and the DRSCABC mnemonic come in.

TRIAGE

You're with a group of people, or at the scene of a natural disaster with mass civilian casualties. You need to decide who to treat first. That's what triage is: a way of having clear priorities. It sometimes involves making some ruthless and difficult calls.

In such situations, it helps to have a simple system of triage categories. There are lots of ways of doing this, but I would recommend placing patients in one of the following four categories:

- **Priority 1:** a critical patient who needs immediate care
- **Priority 2:** a patient who has serious wounds but whose care can be delayed until the critical patients have been dealt with
- **Priority 3:** a patient with minor injuries who needs help but not urgently
- **Priority 4:** a patient who is dead or expected to die

Priority 1 is the most urgent. Priority 4 is the least urgent.

DRSCABC

The traditional triage mnemonic many first responders used was either BBBB (Breathing, Bleeding, Breaks and Burns) or ABCD (Airways, Breathing, Circulation, Disability).

I always liked Breathing, Bleeding, Breaks and Burns as it rolled off the tongue easily in a crisis.

But a more current mnemonic I now use is DRSCABC. This may be harder to remember, but if you can get it into your head it's a more targeted and effective list of priorities to follow.

I remember it this way: 'Don't Rush, Stop & Consider: ABC.'

However you remember it, the principles are proven to save lives.

(Note that this mnemonic DRSCABC is adapted from DRSABC. The extra C, for the likelihood of a catastrophic bleed, reflects the recent experiences of the military in Afghanistan.)

DRSCABC stands for:

- **D**anger
- **R**esponse
- **S**hout for help
- **C**atastrophic bleed
- **A**irway
- **B**reathing
- **C**irculation and signs of life

DANGER

If someone needs your help, you need to check that you're not going to put yourself in unrecoverable danger by going to them, otherwise you may risk making the situation worse for you, the victim and others.

RESPONSE

Your casualty may appear unconscious. Do they respond when you ask them 'Are you all right?' or 'Can you open your eyes?' Does pinching their ear or gently shaking their shoulders elicit a response? If not, your patient requires more immediate attention than somebody who is responsive – unless they are Priority 4 (dead or expected to die).

Remember: the silent, non-breathing casualty is going to need life-saving intervention much sooner than the screaming casualty whose airway is clearly working.

SHOUT FOR HELP

By voice, or by radio or phone, or ask someone in the vicinity to get help – but be prepared for the fact that in an extreme environment neither equipment nor help may be available.

CATASTROPHIC BLEED

We're not talking a nasty cut. We're talking: have they had their arm ripped off? Are they pumping out bright red arterial blood at such a rate that they will die if it's not stopped? You must learn to recognize and treat a catastrophic bleed. If you don't, the victim can die in minutes. See the pages 410–15 for what to do.

AIRWAY

Is their airway open and clear? If not, you need to remove whatever is obstructing it. If the patient is unresponsive, tilt their head back and lift their chin to open the airway. Then check again. You might need to physically clear an airway with your finger if their tongue has fallen back into the airway or vomit is blocking it.

BREATHING

Is the casualty breathing? You can check by putting your ear next to their mouth to hear and feel breaths, and by watching for the rise and fall of their chest. If they're not breathing you need to start giving them CPR (see pages 417–21).

CIRCULATION AND SIGNS OF LIFE

Check for a pulse by putting your index and middle fingers against the side of their neck, on the inside of their wrist or under the armpits. Sometimes it's hard to find a pulse, for example if a casualty is very weak or has been in cold water. Even experienced medics can find it difficult. If there is no sign of a pulse, check for other signs of life: eye movement, venous return (squeeze a fingernail – if it goes white then red again, it means circulation is present), warmth. If you can detect none of these, you need to start giving CPR.

THE RECOVERY POSITION

If there is no catastrophic bleed and the casualty is breathing and has a pulse but is still unconscious, get them into the recovery position. This will keep the airways clear and allow fluid, mucus and vomit to drain away while they're unconscious.

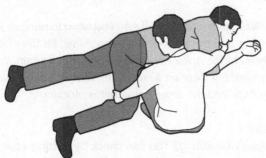

Always monitor a casualty who is in the recovery position. Or ask someone else to do this if you are the only one on site who knows what to do elsewhere. If they stop breathing, roll them on to their back and start CPR.

If you can't get a casualty into the classic recovery position for any reason (such as their legs being trapped), try to ensure that:

- the chest and abdomen are off the ground
- the airway is open and maintained
- they are in a stable position

So if, for example, an unconscious casualty is sitting in a car with their legs pinned, you should sit in the seat behind them, reach out to hold their head with two hands, gently tilt their head back and keep hold of it until help arrives.

K.I.S.S.
(KEEP IT SIMPLE, STUPID!)

Remember your own safety first, so as not
to add to the casualty list.

Know the four triage priorities: clear prioritization
of victims saves lives.

Be prepared to make tough calls: a casualty who is
expected to die has the lowest priority.

DRSCABC.

Remember: the silent, non-breathing
casualty needs your help first.

HOW TO

DEAL

WITH A

CATASTROPHIC

BLEED

f you've read the section on triage, you'll know how important it is to deal with a catastrophic bleed quickly.

Don't get confused. A catastrophic bleed is not necessarily simply a wound that looks meaty and nasty, or a cut so deep that it goes down to the bone. A catastrophic bleed, or haemorrhage, is:

- severe
- sustained
- uncontrolled
- likely to cause death within minutes

Normally it is an arterial bleed. Arterial blood is bright red, frothy and often pumps out in the same rhythm as your heartbeat. A venous bleed – from a vein – is normally slower and the blood is darker. It can still be severe, however.

Understand this: the advice I'm about to give you is not for lesser cuts and bleeds, or for the average first aid situation where professional medical help can immediately be sought. If you can contact the emergency services, you should always do that first. But in a jungle, or a war zone, or somewhere far from civilization where it's clear that a catastrophic bleed will lead to death within minutes and there is no way of getting any other immediate help, here's what you need to do. It is survival medicine at its most raw.

Your treatment of a catastrophic bleed will depend on which part of the body is injured.

AN OPEN CHEST WOUND

If something punctures the lung, or the gap between the lung and the ribs, your first priority should be to apply a chest dressing. If you've read page 22, you'll know that I always carry an Asherman Chest Seal. It's the perfect dressing for a sucking chest wound because it has a valve on it that lets air out but not in – important, because if air gets trapped in the chest cavity there can be a build-up of pressure causing the lung to collapse.

If you have no dressings or chest seals, gaffer tape or a plastic bag sealed over the hole will do a job. Just be aware that you may need to relieve the pressure inside the chest cavity by leaving part of the hole open, or by removing the gaffer tape if you think the pressure is building up. Signs of this happening include:

- acute shortness of breath
- bulging veins in the neck
- blue lips or extremities
- one side of the chest looking bigger than the other

A HEAD, NECK, GROIN OR TORSO WOUND

If there is a catastrophic bleed to one of these areas, you will need to apply direct pressure.

If you have a haemostatic dressing (like the Quikclot dressing I keep in my first aid kit – see page 21) you need to pack it tightly into the wound. If you only have a regular field dressing, use that. If not, rip off your T-shirt or get a towel or rag from your pack and get that into the wound.

Now you need to apply as much direct pressure as you can. Keep it going for at least five minutes, until the bleeding starts to become controlled. When it does, keep the dressing inside the wound (this is key), wrap it with another dressing and do whatever is necessary to get the casualty to safety.

A LIMB WOUND

In a survival situation, a catastrophic bleed from a limb – especially if that limb has been amputated or severed – will likely require a tourniquet. This is a constricting bandage that stops the flow of blood.

The use of tourniquets is controversial in the first aid world, and not without reason. You don't want to be using them where they're not really needed, because if you do there's a genuine risk that your patient will end up having to have their limb amputated. Certainly if a wound is not catastrophic, direct pressure is the way to go. In the survival world, however, tourniquets are not controversial at all. When there's no hope of immediate medical care, and it's a life-or-death situation, a tourniquet is the only way to stop a catastrophic bleed from a limb.

I carry a tourniquet in my pack. If you don't have one, you'll need to improvise. You could use a belt, or some elastic from your coat, or bits of material from your shirt tied together, or a jungle vine. Whatever you use, this is what you need to remember:

- If the wound is in the upper arm or thigh, the tourniquet should be applied at least 5cm above it.
- If the wound is below the elbow or knee, the tourniquet should be applied just above the elbow or knee joint.
- The tourniquet needs to be as tight as you can make it.

And when I say tight, I mean *tight*. If it's too loose, it can actually make the bleeding worse. If the casualty starts screaming with pain because it hurts, you've got it about right. If you're using an improvised tourniquet, your fingers probably won't be strong enough, and you'll need to use a stick to get some purchase, like this:

Once the tourniquet is tight enough, any bright red arterial blood should stop flowing, although there may still be some oozing dark red venous blood. If there is still copious bleeding, apply direct pressure or a haemostatic dressing to the bleeding area. Just because the arterial bleeding has stopped, however, don't remove the tourniquet as this might allow the wound to start bleeding again.

You may need to tighten the tourniquet after it's been on for a bit, as it can cause the limb to become less swollen.

A tourniquet stops the circulation. If you're using it for a wound that is not a sever or an amputation, release it slightly every fifteen minutes to get the blood flowing to the limb again.

K.I.S.S.
(KEEP IT SIMPLE, STUPID!)

A catastrophic bleed is severe, sustained and uncontrolled.

Bright red arterial blood is serious – stuff it and stop it however you can.

Open chest injury: apply a chest dressing.

Head, neck or torso injury: apply direct pressure.

Catastrophic limb injury: apply a tourniquet.

HOW TO

PERFORM

LIFE-SAVING

CPR

CPR stands for cardiopulmonary resuscitation. You perform CPR when you have a non-breathing casualty. It comprises two actions: chest compressions and rescue breaths. The process is slightly different for adults and children.

I'm going to give you the details below, but you should know that the guidelines often get updated. That said, in a survival situation you can't worry too much about the details. The basic principle is this: you need to get the casualty's heart pumping using the heel of your hand against their chest, and you need to get some air in their lungs by pinching their nose and blowing into their mouth.

It's harder than it sounds, and it can be ugly – people often break their casualty's ribs doing it. But that's a secondary concern next to the life-saving priority of getting oxygen into the body.

It's amazing how long you can keep somebody alive doing this. I've heard people say that if the casualty still isn't breathing after ten minutes then they're dead, or at least brain dead, and you should leave them. Don't believe it. My mate's friend received CPR for a full thirty-five minutes before the medics were able to defibrillate him. Three weeks later he was back home without any ill effects. The take-home message is: in a life-saving situation, never, ever give up.

CPR FOR ADULTS

With your patient on their back, kneel down next to them, level with their chest. Now perform thirty chest compressions. To do this:

1. Put the heel of one hand on the centre of the chest where the two halves of the ribcage meet. Put your other hand on top and interlock your fingers so that you're only performing the compressions with the heel of your hand. Keep your fingers clear of the chest.
2. Kneel over the chest with your arms straight and press down so the ribcage sinks by about 5–6cm.
3. Release the pressure without taking your hands away.
4. Repeat this thirty times, at the rate of two compressions a second. Most people do this by remembering the beat to the Bee Gees song 'Staying Alive'.

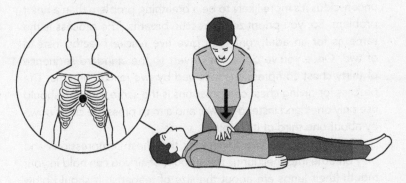

You now need to perform two rescue breaths. To do this:

1. Tilt back the casualty's head and lift their chin.
2. Pinch their nose.
3. Take a normal breath and put your mouth over the casualty's mouth. Make sure there are no air gaps.
4. Blow into the lungs until the chest rises – it should take about a second.
5. Take the mouth away and let the chest fall.
6. Repeat.

Continue this sequence of chest compressions and rescue breaths until the casualty starts to breathe normally, or help arrives. If the patient does start breathing again, get them into the recovery position (see pages 407–8).

If you suspect a breathing issue, such as drowning or asphyxiation, prioritize the rescue breaths. Give five of these, then revert to the thirty chest compressions/two rescue breaths regime.

With all rescue breaths, breathe slowly and constantly and try to gauge the casualty's lung size with your own. If you put too much air in, or do it too fast, some will go into the stomach and the casualty will vomit (probably into your mouth).

CPR FOR CHILDREN

The process is slightly different for children because if a child is unconscious it's more likely to be a breathing problem than a heart problem. So, you prioritize the rescue breaths. The process is the same as for an adult, only you give five rescue breaths instead of two. Once you've done this, revert to the standard sequence of thirty chest compressions followed by two rescue breaths. The process for giving chest compressions is the same, but you should use only one hand instead of two, and aim to press the chest down by about one third of its depth.

For babies, use two fingers for the chest compressions and only breathe into their lungs the amount of air you can hold in your mouth (their lungs are about the size of teabags). It should blow for about a second, which should be enough to see the baby's chest rise.

AGONAL BREATHING

This happens when an unconscious casualty appears to be breathing even though their heart has stopped. It means they are dying. Signs of agonal breathing are.

- short, sharp breaths, with the mouth fixed open
- gasping, gurgling sounds
- no rising and falling of the chest
- no pulse

If you suspect agonal breathing, start CPR immediately.

K.I.S.S.
(KEEP IT SIMPLE, STUPID!)

Non-breathing or agonal breathing casualty: perform CPR.

Adults: thirty chest compressions, two rescue breaths, repeat.

Children: five rescue breaths, then repeat the sequence of thirty chest compressions and two rescue breaths.

Don't give up: you can keep someone alive
for a long time doing this.

HOW TO
STITCH
A WOUND

Wounds that don't emit catastrophic amounts of blood have a lower triage priority. In a survival situation, however, they can still kill. They might not be spurting out a pint of blood a minute, but they can still cause heavy, life-threatening blood loss. They can easily become infected, too, which is in itself a potentially lethal condition. And if the wound is bad and deep, it can stop you moving and getting out of your survival situation alive.

If you think you can get to a hospital within twenty-four hours, I'd advise getting a bandage round any non-catastrophic bleed. But if you can't, knowing how to close up a cut can be a useful skill. It'll help it heal naturally, it can get you back on your feet, and it will stop the wound getting worse if you have no choice but to start moving again. I'd consider this especially if:

- a cut is more than an inch long
- you can't get the sides of the wound to stay together
- you can't get the bleeding to stop by applying heavy pressure
- you can see muscle, bone or joint through a gaping cut

ALTERNATIVES TO STITCHING

Stitching a wound is generally a last resort. In any case, you often won't have the material required to do it. You have other alternatives. If you have steri-strips, use them. If you have gaffer tape in your pack, you can use it to close the wound up and it will often get the job done. In A&E departments they use a special medical glue to stick wounds together. Regular superglue does a similar job, and it's reasonably sterile. This is often a better choice than stitching.

WHAT TO USE

If you have a proper suture kit, use that. If not, you may have to improvise. Dental floss and fishing equipment have both saved lives in the field. These improvised materials are not ideal, but if that's all you have in a life-or-death situation, that's what you'll have to use.

You'll need a needle. Try to sterilize it before you start, using boiling water, a flame or alcohol.

HOW TO DO IT

Your first step is to clean the wound if at all possible. If you don't, there's a high chance of infection. Do this with clean water, flooding the wound and washing out any debris from its centre. If you have a syringe, use this as an irrigating device. It allows you to get water directly into the wound, and helps conserve water if it's scarce.

I could give you a list of different stitches and techniques, but essentially there are two kinds: interrupted and continuous. Interrupted stitches are individual stitches each of which needs tying – time-consuming and difficult. Continuous stitches use the same line without cutting it. This is the fastest and easiest option. I'd recommend a simple over-and-over stitching technique:

Your stitches don't need to be neat or intricate. You're just aiming to keep the wound closed. People generally advise having them about an eighth of an inch apart. Personally, in a survival situation I'd use as few stitches as necessary to hold the wound together. The time may come when you want to remove the stitches yourself with your knife. The fewer there are, the easier and less painful this will be.

Don't make your stitches too deep. You're aiming for the layer just below the skin but no deeper. You'll need to use your judgement depending on what part of the body you're stitching, as skin thickness is not uniform.

Once you've finished your stitching, tie the line at either end with simple overhand knots. Now cover the wound with a clean bandage if you have one, to keep it clear of infection. Change the bandage at least once a day if you can.

K.I.S.S.
(KEEP IT SIMPLE, STUPID!)

You can use gaffer tape or superglue if you have no steri-strips or stitches.

Sterilize your needle.

Use a simple over-and-over continuous stitch, not too deep.

Bandage the wound once it's stitched.

HOW TO

TREAT

A BURN

Burns are categorized into three types:

- *Superficial* burns, which affect the epidermis, or outer layer of skin.
- *Partial-thickness* burns, which affect the epidermis and the dermis (the lower layer of skin).
- *Full-thickness* burns, which affect the tissues beneath the skin. They are the most dangerous, but can be the least painful because they can destroy the nerve endings.

The immediate treatment for any kind of burn is as follows:

- Stop the burning process. This might mean dousing a fire, smothering burning clothes or simply removing the person from the source of the burning.

- Remove any clothing or jewellery in the vicinity of the burned area, unless it's stuck to the burned skin, in which case leave it.

- Place the burn under cool running water for at least ten minutes. If this isn't possible, use whatever cool liquids you have.

- Wrap the burn in clingfilm, or place a plastic bag over it. This not only keeps the injury moist (note that cool running water can still be applied over the clingfilm), but the clingfilm also acts to keep the wound clean. If these items are not available, use a sterile, non-fluffy dressing. Don't use anything adhesive – no plasters or tape.

- Don't use any creams or ointments, ignore the old wives' tale about rubbing with butter, and don't pop any blisters that form.

- Get some water inside yourself or the casualty. Burns cause fluid loss.

- If the face is burned, sit up: this will reduce swelling.

You need to get to a hospital asap if:

- it's a large superficial or partial-thickness burn – bigger than a hand size

- it's a full-thickness burn (you can identify these by white or charred skin)
- the burn is on the face, over a major joint, or on the hands, feet or genitals
- the burn blisters badly

If you're in contact with **medics**, they may want to know the extent of the burns. It's worth knowing the 'rule of hands': the palm of your hand covers about I per cent of the human body, so a burn area the size of five palms is a 5 per cent burn.

CHEMICAL BURNS

If you are treating a chemical burn, you should:

- remove any contaminated clothing from you or the casualty
- brush off any dry chemical residue
- use cool running water to flood the burned area and remove any traces of the chemical
- not try to neutralize the burn using an acid or alkali, which can make things worse

ELECTRICAL BURNS

The immediate treatment for electrical burns is the same as for chemical burns. However, before administering it you need to remove the casualty from contact with the electrical source. Kill the electrical source at the main power switch if you can. If you can't, you need to separate the casualty from the source. Do this by standing on a dry surface – a rubber doormat is ideal. You mustn't be standing in, or near, water. Use a long wooden object like a broom handle to separate the person from the source. Don't use anything wet, or made of metal.

Before you tend to the burns, check your casualty's breathing and heart rate. If either has stopped, administer CPR (see pages 417–21).

If you need to treat the burns, be aware that they may exist both where the electricity entered and where it exited the body.

Electricity can cause internal burning, too. Even if there are no signs of external burning, get the casualty to hospital immediately.

K.I.S.S.
(KEEP IT SIMPLE, STUPID!)

Remove clothing or jewellery, but not if it's stuck to the skin.

Flood the burned area with continuous cool water and keep it there for at least ten minutes.

Sometimes the most serious burns hurt the least, because of nerve damage.

Use insulating materials to remove an electrical burns casualty from the electrical source.

HOW TO

FIELD-DRESS

A BROKEN

BONE

Broken bones can be excruciatingly painful, but they're not necessarily life-threatening in themselves. They become so either when they cause significant internal or external bleeding, or when they hinder your or your companion's ability to escape in a survival situation.

Your immediate priority will be to establish if a fracture is open or closed. An open fracture is easy to diagnose because it means that the bone is jutting out through the skin. A closed fracture is more difficult because you need an X-ray. If you suspect a closed fracture, you should treat it as such. If you don't, shards of bone can easily tear into arteries and veins, causing life-threatening internal bleeding.

In a survival situation, your immediate priorities are to treat the bleed if there is one, then to immobilize the joint to increase the chances of you or your casualty getting to safety without the situation worsening.

OPEN FRACTURES

These can look dramatic and messy. There can be a lot of blood and your first priority must be to treat the bleed (see pages 411–15). The main problem with an open fracture is the risk of infection. If this sets into the bone, you've got a major problem. You must keep the wound clean. Do this by flooding it regularly with clean water. You shouldn't wrap a bandage around the wound, because of the need to irrigate it regularly. But you can cover it, just to keep it as uncontaminated as possible.

GENTLE TRACTION

In general you should manipulate a broken limb as little as possible. However, if a casualty is experiencing extreme pain and there is no hope of proper medical care, you can try applying gentle traction to bring the broken ends of a bone back in line with each other. You do this by pulling the broken limb firmly but slowly so that it slips back into place. Get a good grip (you may have to pull for several seconds or even minutes), and pull until the ends of the bone are aligned. *Don't* yank it.

SPLINTS

If it's a very bad break, you might want to splint the arm too. The purpose of a splint is to stabilize the fractured limb. It's normally used for leg breaks, but you can splint an arm if it's a bad break or you don't have sling material.

You'll need to use whatever comes to hand: a plank of wood, a straight tree branch, a pole. Rolled-up newspapers are stiff enough to do the job, and soldiers have been known to use unloaded rifles for this purpose.

If you're using a splint:

- Don't try to realign any broken bones.
- Make sure the splint extends above *and* below the joints

on either side of the break. So, for a broken shin, extend the splint past the ankle and the knee.

- Tie or gaffer-tape the splint to the limb, but not too tightly – you don't want to constrict the blood flow.

Leg breaks don't necessarily require splints. You can strap or tape the broken leg to the good one, which will do the job for you. If you do this, it's important to pad out any areas where bone meets bone, otherwise you're going to get circulation problems. Socks, pants, hats – anything will do.

SLINGS

A sling is a quick, easy way of immobilizing a break in either the upper or lower arm, or the collar bone, using a triangular bandage.

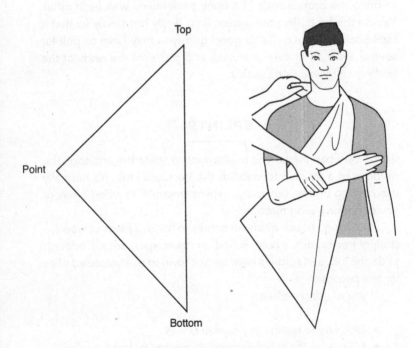

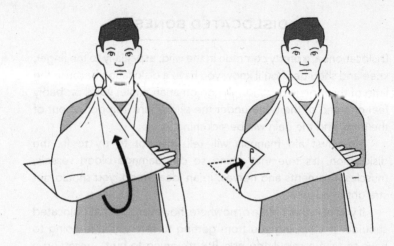

If you don't have a triangular bandage, improvise by cutting up an item of clothing. Alternatively, use a belt looped into a figure of eight, like this:

Or you can use duct tape if you have it in your pack – it's strong enough and flexible enough for the job. The best way to do this is to fold the tape so that none of the adhesive is exposed and so doesn't stick to your skin.

Likewise you can use clingfilm to improvise a sling.

DISLOCATED BONES

Dislocations are pretty common in the wild, especially to the finger, knee and shoulder. You'll know you have a dislocation because the limb of the digit in question will go off at an angle, you'll probably feel one end of the bone under the skin where it's popped out of the joint, and the pain will be sickening.

Most first aid manuals will tell you not to try to fix the dislocation. It's true that doing so can damage blood vessels, muscles, ligaments and nerves. In an ideal world, you should get straight to hospital.

If you're in the middle of nowhere, however, and that dislocated shoulder is stopping you from getting to safety, you're going to have to take a calculated risk. But it's going to hurt – resetting a dislocated shoulder, for example, is one of the most painful things you can do.

If you do reset a dislocated bone, go carefully with it as it's now more prone to dislocating itself in the future.

FINGER

If you can ice the joint in very cold water first, do so to help with the pain. Now, wrap the finger in a piece of cloth or clothing so you can get a good grip. If possible, have someone else hold the wrist. Pull the finger firmly. If it doesn't click into place, pull a little harder, but don't jerk it. You're trying to stretch the tendons just enough that the finger slips back into place.

KNEE

It's best to have someone to help you with this. You need your hip to be flexed, so you should either sit down, or lie down and raise your thigh at about a 30° angle. Now, get someone to slowly raise your knee. Manipulate the knee cap as they do so. Do it slowly and don't force it or you'll make matters much worse.

SHOULDER

A friend of mine has had forty dislocations of one shoulder and five of the other. To put his back in he would go on all fours, try to relax and then gently shake his arm and shoulder. Sometimes that worked. Other times he would have to gently rotate his arm through 180° so that it found its way back in, the same way it came out. Do this slowly, and never force it. It'll be just about the most painful thing you've ever done.

Alternatively, lie on your back with the dislocated arm pointing away from your body at an angle of 90°. Get a companion to grab your wrist and pull, slowly but firmly, so that you create gentle traction. They shouldn't make any sudden movements. If necessary, your companion can put their foot against your torso to get purchase. Their aim is to get the top of your humerus bone to slip back into its socket. You'll hear it when it happens.

Once you've relocated a shoulder, the pain will reduce but it will still be very sore and weak. Get it in a sling and keep it immobile if possible.

K.I.S.S.
(KEEP IT SIMPLE, STUPID!)

Open fracture: treat the bleed before the break.

Immobilize a broken bone using a sling or a splint.

Don't constrict the blood flow.

Reset dislocated bones only if the injury is preventing you from getting to safety, and expect it to hurt.

HOW TO

TREAT

HYPERTHERMIA,

HYPOTHERMIA

AND

FROSTBITE

Your core body temperature should be between 36°C and 37°C. One degree higher or lower than these values and you're too hot or cold. Two degrees higher or lower and you're hyper- or hypothermic. Both conditions are common in survival situations. They are life-threatening and require immediate attention. I have witnessed both of these first-hand on many occasions and they need to be dealt with rapidly or you can easily lose people.

HYPERTHERMIA

Know the warning signs:

- Heavy sweat. This means your body is working overtime to keep itself cool. Even more dangerous, though, is heavy sweating followed by no sweating at all. This is your body going into survival mode and conserving its last reserves of water.
- Thirst. If you're thirsty, you're already dehydrated. In hot climates you need to be drinking constantly. If you've suffered diarrhoea or vomiting, this is even more important.
- Muscle cramps.
- Nausea or dizziness.
- Tiredness and confusion.
- Dark urine.
- High pulse.

If you experience any of these, you need to do two things: drink water, and reduce your core temperature.

WATER

If you're hyperthermic, you need to replace your body's fluids quickly. However, you mustn't gulp water down, as your body might reject it. Take small sips. If you have rehydration salts, use them to replace lost electrolytes in the body. (You can make your own by dissolving a teaspoon of salt in a litre of water.)

CORE TEMPERATURE

Get yourself or your casualty into the shade. Stop them from moving around. If at all possible, wrap them in a cold, wet sheet. Keep the sheet wet by pouring water over it at intervals. If you can't do this, fan them to try to bring their temperature down.

Don't underestimate the importance of this. The body will have lost the ability to regulate its own temperature, so it needs help doing it. Without this help it won't be able to recover and death is a real possibility.

If the core temperature stays high, the casualty might become unconscious and stop breathing, so you'll need to administer CPR (see pages 417–21).

HYPOTHERMIA

Again, know the warning signs:

- Uncontrollable shivering. This is the body trying to warm itself up through movement. If you or your companion suddenly stops shaking, it's the body going into survival mode, shutting down blood flow to the extremities and concentrating on blood flow to the heart, lungs and brain.
- Pallor in the lips.
- Body feeling cold to the touch.
- Lack of coordination, slowness of movement, confusion and a tendency to make silly mistakes.
- Quietness.

If you suspect hypothermia:

- Remove any wet clothes and replace them with dry ones.
- Get the casualty into a sleeping bag. If possible, get into the sleeping bag with them. Shared body heat is good, and flesh-on-flesh contact is better. Don't be shy about this: it can save a life.
- Get plenty of hot, sweet drinks inside the casualty.
- Place hot water bottles or chemical heat packs against the armpits, upper abdomen, neck or groin.
- Encourage the casualty to urinate – the body uses a lot of heat energy keeping urine warm inside the bladder.

There are some things you *shouldn't* do:

- **Don't** immerse the casualty in very hot water. This can force warm blood towards the skin, away from the body's core where it's needed.

- **Don't** warm the casualty's arms and legs with direct heat.
- **Don't** rub or massage the casualty.

If the casualty is severely hypothermic, there's a chance his or her breathing will stop. Be prepared to administer CPR (see pages 417–21).

FROSTBITE

Frostbite happens when your flesh freezes. The skin goes black as the flesh starts to die. In acute cases, the frostbitten area will have to be amputated.

The first stage of frostbite is frostnip. This normally occurs on your fingers and toes and you can tell it's happening because they start to turn white and numb. If you suspect frostnip, you need to get the affected area out of the wind, rain or snow and gently re-warm it. The best way to do this is under your – or someone else's – armpits or in the groin. You can squeeze and manipulate the flesh to get the circulation going.

Frostbite is more serious. You can distinguish it from frostnip because the flesh will feel very solid to the touch. If that happens, you need slowly to re-warm the flesh in lukewarm – not hot – water. If this is not available, try using your warm breath on the frostbitten area. You'll know the frostbite has been dealt with when the flesh is red and swollen. Don't rub it and don't re-warm over direct heat: there will be no sensation in the affected area, so you risk burning yourself.

You should never re-warm frostbite if there's a chance the affected area will refreeze as this can damage the flesh beyond repair. Better to keep it frostbitten until you can get to safety.

Be warned: re-warming frostbite is one of the most excruciatingly painful things you'll ever do. I've seen people with frostbite try to put their hands back out into –10°C conditions because of the pain of having them at room temperature.

K.I.S.S.

(KEEP IT SIMPLE, STUPID!)

Hyperthermia: slow sips of water, shade,
wet towels round the body.

Hypothermia: warm drinks, shared body heat, avoid
very hot water or direct heat to the limbs.

Frostnip: slow re-warming and gentle manipulation.

Frostbite: slow re-warming, but keep the affected area
frostbitten if there's a chance of refreezing.

INDEX